Computational Fluid Dynamics - Recent Advances, New Perspectives and Applications

Edited by Guozhao Ji and Jingliang Dong

Published in London, United Kingdom

Computational Fluid Dynamics - Recent Advances, New Perspectives and Applications
http://dx.doi.org/10.5772/intechopen.104020
Edited by Guozhao Ji and Jingliang Dong

Contributors
Valiyollah Ghazanfari, Stefan Seidel, Rüdiger W. Maschke, Cedric Schirmer, Lia Rossi, Regine Eibl, Dieter Eibl, Yajiao Liu, Junhu Dang, Xiaosen Dai, Hao Luo, Tiecheng Zhao, Yipeng Wu, Lingquan Li, Jialin Lou, Hajime Ikeda, Akinori Sakabe, Shingo Hamada, Mitsuhiro Shitota, Takashi Kobayashi, Guozhao Ji, Jingliang Dong, Meng Zhang, Yongming Lu

First published in London, United Kingdom, 2023 by IntechOpen
IntechOpen is the global imprint of INTECHOPEN LIMITED, registered in England and Wales, registration number: 11086078, 5 Princes Gate Court, London, SW7 2QJ, United Kingdom

British Library Cataloguing-in-Publication Data
A catalogue record for this book is available from the British Library

Additional hard and PDF copies can be obtained from orders@intechopen.com

Computational Fluid Dynamics - Recent Advances, New Perspectives and Applications
Edited by Guozhao Ji and Jingliang Dong
p. cm.
Print ISBN 978-1-80355-654-3
Online ISBN 978-1-80355-655-0
eBook (PDF) ISBN 978-1-80355-656-7

For EU product safety concerns:
IN TECH d.o.o., Prolaz Marije Krucifikse Kozulić 3, 51000 Rijeka, Croatia,
info@intechopen.com or visit our website at intechopen.com.

Meet the editors

Dr. Guozhao Ji received his Ph.D. in Chemical Engineering from the University of Queensland, Australia, in 2014. He completed his MS and BS in Mechanical Engineering at Northeastern University, People's Republic of China, in 2010 and 2008, respectively. Currently, Dr. Ji is an associate professor at the School of Environmental Science and Technology, Dalian University of Technology, China. His research interests include computational fluid dynamic applications in chemical engineering processes, modeling of gas transport in microporous materials, gas separation by inorganic membranes, high-temperature CO_2 capture, solid waste gasification, and kinetic modeling of thermochemical conversions. He has authored more than 100 refereed journal publications, 2 books, and 3 book chapters.

Jingliang Dong, Ph.D., is a lecturer at the First Year College of Victoria University, Australia, and a researcher at the esteemed Institute for Sustainable Industries and Liveable Cities (ISILC) at the same institution. He obtained his Ph.D. in Biomechanical Engineering from RMIT University, Australia, in 2015. After graduating, he worked as a postdoctoral fellow at Chongqing University, China, and a research fellow at RMIT University. In 2020, he was awarded the prestigious Discovery Early Career Researcher Award (DECRA), along with funding, by the Australian Research Council (ARC) to assist him in establishing and leading a research team to conduct particle inhalation exposure risk analysis and cross-species data extrapolation between animal–human airways. In addition, he also extended his research interests to include inhalation therapy. He targeted drug delivery, the hemodynamic study of cardiovascular arteries, heat and mass transfer in fluid machinery, and air pollution in urban street canyons and tunnels. Dr. Dong has led three special journal issues and authored over sixty journal articles, fifteen conference proceedings, and two book chapters.

Contents

Preface

Computational Fluid Dynamics (CFD) is a rapidly evolving field that has seen remarkable growth in recent years. The application of numerical methods and algorithms to study fluid flow has revolutionized our understanding of many phenomena in fluid mechanics, such as turbulence, multiphase flow, and heat transfer, thermal reactions. As such, CFD has become an essential tool for engineers, physicists, and researchers in various fields.

CFD simulation is a transdisciplinary technique across fluid mechanics, mathematical algorithms, and computer science that offers a cost-effective alternative to experimental tests. It can accurately reproduce real-world flow phenomena, capture key information for analysis, and has a wide range of applications in fields such as mechanical engineering, chemical engineering, environmental engineering, and thermal engineering. With the aid of high-performance computers, CFD has made significant progress in recent decades, allowing for even greater accuracy and efficiency in solving practical problems.

This edited book, *Computational Fluid Dynamics – Recent Advances, New Perspectives and Applications*, presents a collection of articles that showcase the latest developments in CFD. The book covers a wide range of topics, from the fundamental principles of fluid mechanics and numerical methods to advanced applications in various engineering and scientific fields. The chapters are contributed by leading experts from academia, industry, and research institutions, providing a comprehensive overview of the current state of the art in CFD. The book is composed of seven chapters. Chapter 1, "The Basic Theory of CFD Governing Equations and the Numerical Solution Methods for Reactive Flows", provides an introduction to the basics of CFD, including governing equations and numerical methods, with a focus on reactive flows. Chapter 2, "Hyperbolic Navier-Stokes with Reconstructed Discontinuous Galerkin Method", presents the Galerkin method as an attractive and viable option for discretizing governing equations. Chapter 3, "Computational Fluid Dynamics for Advanced Characterisation of Bioreactors Used in the Biopharmaceutical Industry – Part I: Literature Review," reviews the use of CFD for process characterisation and determining process engineering parameters in the biopharmaceutical industry. Chapter 4, "Computational Fluid Dynamics for Advanced Characterisation of Bioreactors Used in the Biopharmaceutical Industry – Part II: Case Studies," presents case studies that demonstrate how CFD can be used to determine the process characterization of mechanically driven bioreactors in the biopharmaceutical industry and validate the results experimentally. In Chapter 5, "The Effect of Al_2O_3 Concentration in Annular Fuels for a Typical VVER-1000 Core", the effect of Al_2O_3 with varying volume percentages in annular fuels for a typical VVER-1000 core is investigated using a CFD modeling approach. Chapter 6, "CFD Application in Ground Source Heat Pump System", discusses the use of CFD to predict the heat transfer characteristics of the ground heat exchanger (GHE), a critical component of ground source heat pump (GSHP) systems. Finally, Chapter 7, "Application of CFD to Prediction of

Heat Exchanger Temperature and Indoor Airflow Control Simulation in Room Air Conditioner Development", describes two CFD models used to investigate the performance of room air conditioners in detail, focusing on the fin-and-tube heat exchanger and the operation of the fan.

Although the different applications involved are interdisciplinary in nature, there is a common need to identify the distribution of fluid dynamic parameters, such as velocity distribution, pressure distribution, temperature distribution, and detect their effects. The advances described by the participating authors have greatly contributed towards this goal. During the preparation of this book, all the participating authors spent significant efforts in composing the chapters with their extraordinary knowledge and high motivation and performed serious revisions where needed. The publication of this book would not have been possible without their ongoing support. The time that they have taken away from their busy schedules to contribute to this book was valuable and greatly appreciated. Also, we are thankful to Mr. Dominik Samardzija, Ms. Jasna Bozic, and Ms. Lucija Tomicic at IntechOpen who helped us throughout every step of the publishing process.

We hope that this book will be a valuable resource for students, researchers, and practitioners interested in the latest developments in CFD. We would like to thank all the contributors for their excellent work and commitment to this project, as well as the publishers for their support in bringing this book to fruition.

Guozhao Ji
Dalian University of Technology,
Dalian, People's Republic of China

Jingliang Dong
Victoria University,
Melbourne, Australia

Section 1

Theory of Computational Fluid Dynamics

Chapter 1

The Basic Theory of CFD Governing Equations and the Numerical Solution Methods for Reactive Flows

Guozhao Ji, Meng Zhang, Yongming Lu and Jingliang Dong

Abstract

The universal principles of fluid motion are the conservation of mass, momentum and energy. This chapter will introduce the CFD governing equations and describe how the continuity equation, component equation, Navier-Stokes equation and energy equation were derived from the principles above. With the expanding application of CFD simulation technology, some processes such as fluid-involved reactions, adsorption and permeation, which break the conservation of mass, momentum and energy for fluid phase, should be coupled to CFD model. In view of this, this chapter provided the theories about source terms for the mass equation, momentum equation and thermal energy equation. The technology for solving these governing equations remained a challenge for a long period due to the complexity. Thanks to the development of numerical methods, such as the finite difference method and the finite volume method, these equations can be solved and provide reasonable numerical results of flows, heat transfer and reactions. This chapter also demonstrates the basics of these two major numerical techniques

Keywords: governing equations, finite-difference method, finite-volume method, reacting flow, multiphase flow, source term

1. Introduction

Using Computational Fluid Dynamics (CFD) simulation is of great importance in modeling reactive flows because it allows for the prediction and analysis of complex fluid dynamics and chemical reactions in a controlled and cost-effective environment [1]. In fluid reactors, the reaction rate depends on many factors such as concentrations of reactant fluid and product fluid, temperature and turbulence, and these key factors vary with time and space coordinate [2]. Knowing the evolution and distribution of the heat and mass transfer factors is important in reactor design and operating condition optimization. CFD simulations can be used to model and simulate the behavior of reactive flows, such as combustion [3], pyrolysis [4–6], gasification [7] and some other chemical reactions-driven fluid flow [8], and get detailed information on heat and mass transfer inside reactors.

There are many advantages of CFD over other traditional research methods. For example, CFD simulations can predict the behavior of combustion, which can be used to optimize combustion systems and reduce emissions. CFD simulations can also be used to optimize the design of reaction systems, such as gas combustion in turbines, boilers, and engines, pyrolysis or gasification in thermal reactors by predicting the performance of different designs and optimizing the operating conditions for maximum efficiency. Safety analysis can be performed by CFD simulations as well, by predicting the behavior of the system under different conditions and identifying potential risks. A distinct advantage of CFD simulations is that it can provide a cost-effective means of analyzing and optimizing reactive flow systems, as they can be used to simulate and analyze the behavior of the system under a wide range of conditions without the need for expensive physical testing [9].

The results of a CFD simulation are obtained through numerical methods that solve the fluid flow governing equations using a finite difference or a finite volume method. The numerical methods used in CFD simulations involve dividing the fluid domain into a grid or mesh of discrete cells and then solving the equations governing fluid flow and heat transfer at each of these grid points or cells. The numerical methods used in CFD simulations are complex and require high-performance computing resources. The accuracy of the results depends on the quality of the mesh, the numerical methods used and the convergence criteria [10]. Therefore, CFD simulations require a deep understanding of numerical methods with careful validation and verification to ensure that the results are accurate and reliable.

To better understand CFD simulations, learning the basics of CFD theory is essential for a few reasons. CFD basic theory provides a fundamental understanding of the governing equations and numerical methods used in CFD simulations. This understanding is essential for developing accurate and reliable CFD simulations. A thorough understanding of CFD theory allows for the development of more efficient and accurate simulations. This is because it enables the user to select appropriate numerical methods, grid sizes and convergence criteria. CFD simulations can be complex, and errors can arise due to various reasons. Understanding CFD theory allows the user to identify the source of the error and develop solutions to fix it. CFD theory is constantly evolving, and researchers are continually developing new methods and algorithms to improve the accuracy and efficiency of CFD simulations. A strong understanding of CFD theory is essential for contributing to the advancement of the field.

2. Governing equations

As fluid is one of the existing states of matter, it follows the laws of motion, such as mass conservation, momentum conservation and energy conservation. These laws govern the motion and force of fluid, and thus are called governing equations. These equations are generally expressed in differentiation forms.

2.1 Continuity equation

If our object is an infinitely small grid, continuity indicates that the mass of fluid entering this small grid equals the mass of fluid flowing out of this grid plus the mass of fluid accumulating in this grid. Since the continuity equation is derived from this mass conservation concept, it can also be regarded as a mass conservation equation which is expressed as Eq. (1).

$$\frac{\mathrm{d}m}{\mathrm{d}t} = \sum_{\text{in}} \dot{m} - \sum_{\text{out}} \dot{m} \tag{1}$$

where m is the mass of this small grid and is defined as $m = \rho\Delta x\Delta y\Delta z$. The mass entering and exiting the small grid in the Cartesian coordinate is demonstrated in **Figure 1**. In the actual scenario, the flow direction does not have to be the same as **Figure 1** shows. Any reversed flow direction can be considered if a minus sign is in the front of a term. According to **Figure 1**, the mass entering this grid with the flow is

$$\sum_{\text{in}} \dot{m} = \rho u \Delta y \Delta z + \rho v \Delta x \Delta z + \rho w \Delta x \Delta y \tag{2}$$

Based on the Taylor series expansion for the ρu, ρv, and ρw terms by omitting the higher order terms, the mass exiting out of this grid is

$$\sum_{\text{out}} \dot{m} = \left[\rho u + \frac{\partial(\rho u)}{\partial x}\Delta x\right]\Delta y \Delta z + \left[\rho v + \frac{\partial(\rho v)}{\partial y}\Delta y\right]\Delta x \Delta z + \left[\rho w + \frac{\partial(\rho w)}{\partial z}\Delta z\right]\Delta x \Delta y \tag{3}$$

Substituting Eqs. (2) and (3) into Eq. (1), one can obtain Eq. (4).

$$\begin{aligned} \frac{\partial(\rho \Delta x \Delta y \Delta z)}{\partial t} &= \rho u \Delta y \Delta z + \rho v \Delta x \Delta z + \rho w \Delta x \Delta y \\ &- \left[\rho u + \frac{\partial(\rho u)}{\partial x}\Delta x\right]\Delta y \Delta z \\ &- \left[\rho v + \frac{\partial(\rho v)}{\partial y}\Delta y\right]\Delta x \Delta z \\ &- \left[\rho w + \frac{\partial(\rho w)}{\partial z}\Delta z\right]\Delta x \Delta y \end{aligned} \tag{4}$$

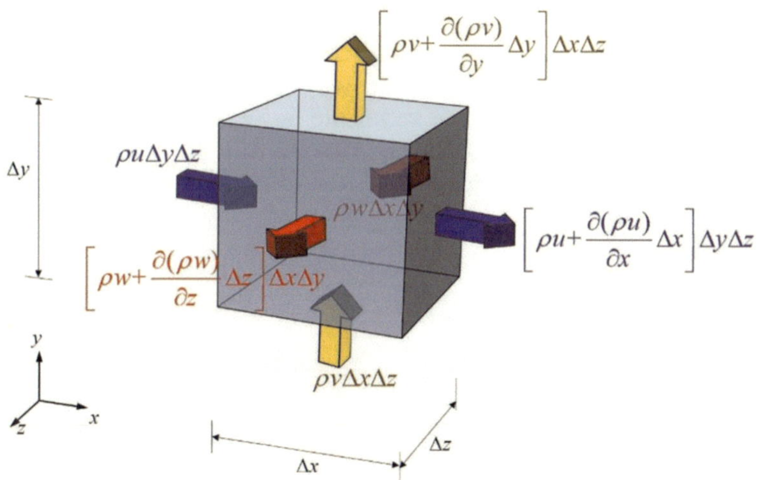

Figure 1.
The schematic of mass conservation in a control volume.

After canceling the identical terms with opposite signs, dividing $\Delta x \Delta y \Delta z$ on both sides, and taking the limit for $\Delta x \to 0$, $\Delta y \to 0$ and $\Delta z \to 0$, we have Eq. (5).

$$\frac{\partial \rho}{\partial t} + \frac{\partial(\rho u)}{\partial x} + \frac{\partial(\rho v)}{\partial y} + \frac{\partial(\rho w)}{\partial z} = 0 \tag{5}$$

Then the continuity equation is derived. Sometimes it is also written in a concise vector form as Eq. (6).

$$\frac{\partial \rho}{\partial t} + \nabla \cdot \left(\rho \vec{u}\right) = 0 \tag{6}$$

2.2 Component balance equation

Even though the continuity equation governs the total mass balance of fluid during its flowing process, for a multi-component flow, the mass balance of each component must also be ensured. The mass transfer of a component is more complex than the total flow. In addition to convective flow, a component could also diffuse under the driving force of the concentration gradient from the high concentration region to the low concentration region.

Figure 2 is a demonstration of diffusive flow. Initially, a chamber is separated by a valve into two sub-chambers with equal volume and pressure. In the left sub-chamber, there are 8 moles of N_2 and 2 moles of CO, while in the right chamber there are 2 moles of N_2 and 8 moles of CO. Once the valve is open, there is no convective flow due to the equal pressure on both sides, but N_2 would diffuse from left to right due to the concentration difference between the left and the right sides. In the meantime, CO would diffuse the other way around owing to the same reason. During this process, there is no convective flow at all, but the mass transfer of each component really occurred. Therefore, when considering the mass balance of a component in a multi-component flow, the diffusive flow or diffusion must be included. Generally, the diffusion follows Fick's law [11]:

$$J_i = D\frac{\mathrm{d}(\rho X_i)}{\mathrm{d}x} \tag{7}$$

where J_i denotes the diffusive flux of component I, D is called diffusivity which is a function of temperature and pressure, and could be evaluated based on Fuller Eq. [12], ρ is the flow density and X_i is the mass fraction of component I in the fluid.

In multi-component flow, the ith component follows the mass balance of Eq. (8)

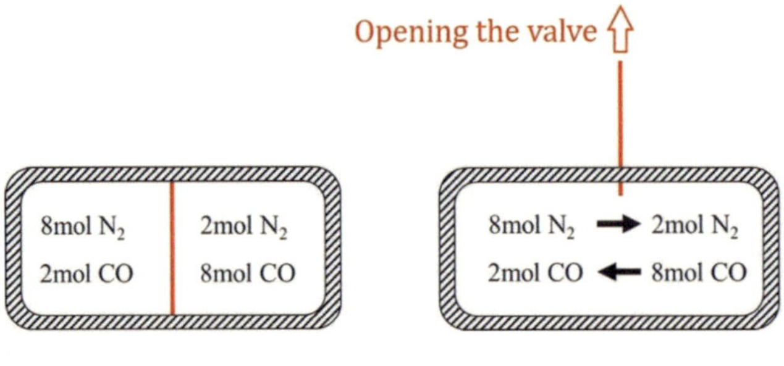

Figure 2.
Demonstration of diffusive flow.

$$\frac{\mathrm{d}m_i}{\mathrm{d}t} = \sum_{\text{in}} \dot{m}_i - \sum_{\text{out}} \dot{m}_i \tag{8}$$

where m_i is the mass of the ith component in a control volume. The ith component entering and exiting the control volume in terms of convective flow in Cartesian coordinate is demonstrated in **Figure 3**. Taking the x direction as an example, the total flow into this control volume from the upwind y-o-z surface is $\rho u\Delta y\Delta z$ (the left blue arrow). If the mass fraction of component I is X_i, the flow of component I into the control volume together with the total convective flow is $X_i\rho u\Delta y\Delta z$. The convective flow of ith component exiting from the downwind y-o-z surface (the right blue arrow) is then expressed by Taylor series expansion by omitting higher-order terms $\{X_i\rho u + [\partial(X_i\rho u)/\partial x]\Delta x\}\Delta y\Delta z$. Likewise, the ith component entering and exiting along y and z directions are $X_i\rho v\Delta x\Delta z$, $X_i\rho w\Delta y\Delta x$, $\{X_i\rho v + [\partial(X_i\rho v)/\partial y]\Delta y\}\Delta x\Delta z$, $\{X_i\rho w + [\partial(X_i\rho w)/\partial z]\Delta z\}\Delta y\Delta x$, respectively.

In addition to the convective flow depicted in **Figure 3**, there is also diffusive flow driven by the concentration gradient (**Figure 4**). According to Fick's law, the diffusive flow of component I entering from the upwind surface of y-o-z is $D[\partial(\rho X_i)/\partial x]\Delta y\Delta z$. Then, the diffusive flow exiting from the opposite surface is $D\{\partial(\rho X_i)/\partial x + [\partial^2(\rho X_i)/\partial x^2]\Delta x\}\Delta y\Delta z$. Likewise, along the y direction, the diffusive flows entering and exiting are $D[\partial(\rho X_i)/\partial y]\Delta x\Delta z$ and $D\{\partial(\rho X_i)/\partial y + [\partial^2(\rho X_i)/\partial y^2]\ \Delta y\}\Delta x\Delta z$. Similarly, along the z direction, the diffusive flow entering and exiting are $D[\partial(\rho X_i)/\partial z]\Delta x\Delta y$ and $D\{\partial(\rho X_i)/\partial z + [\partial^2(\rho X_i)/\partial z^2]\ \Delta z\}\Delta x\Delta y$.

Substituting all the convective flow and diffusive flow expressions into Eq. (8), and canceling out the identical terms with opposite signs leads to Eq. (9).

$$\begin{aligned}\frac{\partial}{\partial t}(\rho X_i\Delta x\Delta y\Delta z) = {} & D\frac{\partial^2(\rho X_i)}{\partial x^2}\Delta x\Delta y\Delta z + D\frac{\partial^2(\rho X_i)}{\partial y^2}\Delta y\Delta x\Delta z + D\frac{\partial^2(\rho X_i)}{\partial z^2}\Delta z\Delta x\Delta y \\ & - \frac{\partial(\rho u X_i)}{\partial x}\Delta x\Delta y\Delta z - \frac{\partial(\rho v X_i)}{\partial y}\Delta y\Delta x\Delta z - \frac{\partial(\rho w X_i)}{\partial z}\Delta z\Delta x\Delta y\end{aligned} \tag{9}$$

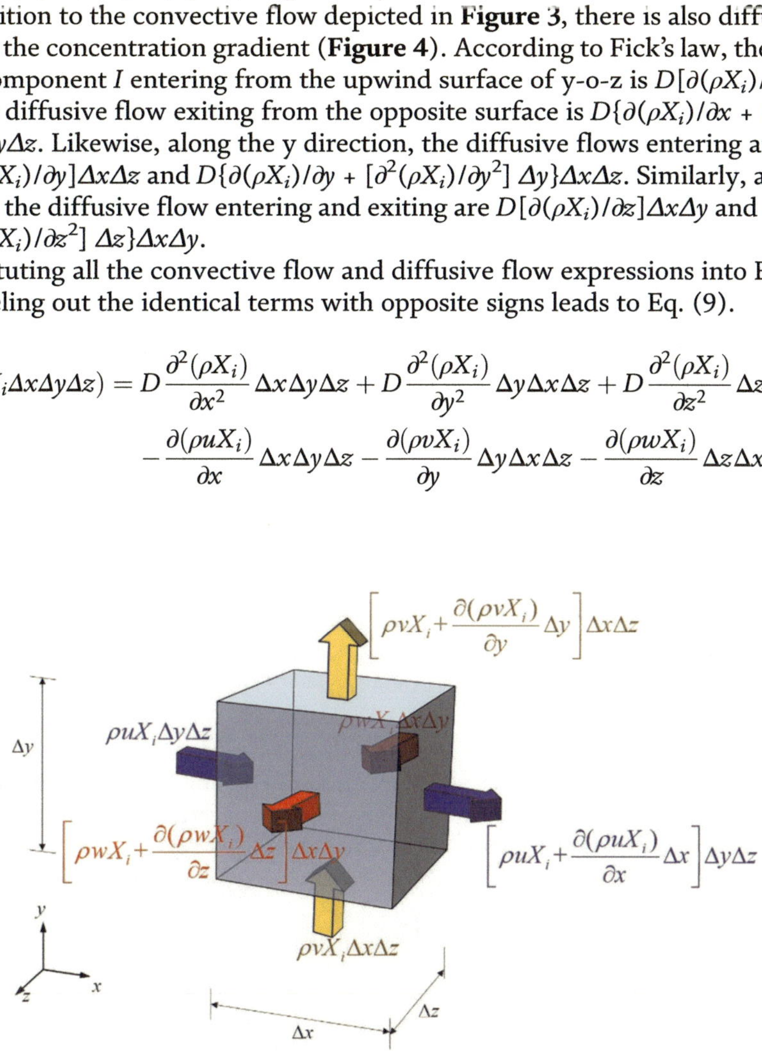

Figure 3.
The schematic of the convective flow of I *component entering and exiting a control volume.*

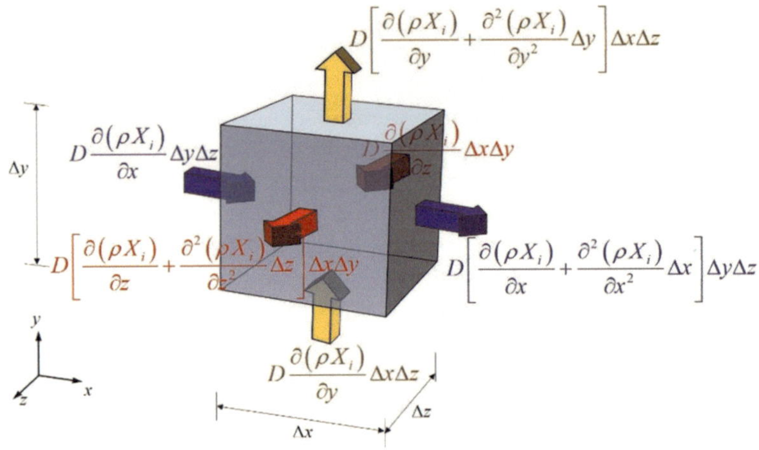

Figure 4.
The schematic of the diffusive flow of I *component entering and exiting a control volume.*

After dividing $\Delta x\Delta y\Delta z$ for both sides, Eq. (9) could be rearranged to Eq. (10)

$$\frac{\partial(\rho X_i)}{\partial t}+\frac{\partial(\rho u X_i)}{\partial x}+\frac{\partial(\rho v X_i)}{\partial y}+\frac{\partial(\rho w X_i)}{\partial z}=D\left[\frac{\partial^2(\rho X_i)}{\partial x^2}+\frac{\partial^2(\rho X_i)}{\partial y^2}+\frac{\partial^2(\rho X_i)}{\partial z^2}\right] \tag{10}$$

which is the final expression of the component balance equation. Component equation is especially important in governing the flow with chemical reactions, because the reasonable prediction of reaction rates depends on accurate information about component distribution in a reactor.

2.3 Momentum equation

Newton's second law is the principle to express the correlation between force and motion of the fluid. Namely, the force acting on an infinitesimal control volume equals the product of its mass and acceleration, which is given as (Eq. (11)).

$$\vec{F}=m\vec{a} \tag{11}$$

Actually, both force and acceleration are vectors and should be expressed by the three components in the x, y and z directions. The acceleration of fluid motion is defined as the total differential with respect to time (Eq. (12)).

$$a_x=\frac{\mathrm{D}u}{\mathrm{D}t} \tag{12}$$

$$a_y=\frac{\mathrm{D}v}{\mathrm{D}t} \tag{13}$$

$$a_z=\frac{\mathrm{D}w}{\mathrm{D}t} \tag{14}$$

Taking the x direction as an example, the acceleration in the x direction is

$$a_x = \frac{\mathrm{D}u}{\mathrm{D}t} = \frac{\partial u}{\partial t} + u\frac{\partial u}{\partial x} + v\frac{\partial u}{\partial y} + w\frac{\partial u}{\partial z} \tag{15}$$

According to Newton's second law, the acceleration in the x direction should equal the quotient of the forces along the x direction divided by mass.

$$a_x = \frac{\sum F_x}{m} \tag{16}$$

Basically, there are two types of forces. Namely, surface forces such as normal stress, shear stress, and body forces such as gravity force and magnetic force. **Figure 5** depicts all the surface forces acting on a control volume along the x direction. Firstly, there is a pair of normal stresses imposed on the left and right surfaces that are indicated as blue arrows. This surface force is caused by internal pressure and velocity gradient in x direction. Secondly, there is a pair of shear stresses on the top and bottom surfaces that are marked as yellow arrows. This surface force is caused by the velocity gradient. Thirdly, a pair of shear stresses are also acting on the front and back surfaces shown as red arrows.

The summation of all the surface forces is

$$\sum F_{\text{surf}} = \frac{\partial \sigma_{xx}}{\partial x}\Delta x\Delta y\Delta z + \frac{\partial \tau_{yx}}{\partial y}\Delta x\Delta y\Delta z + \frac{\partial \tau_{zx}}{\partial z}\Delta x\Delta y\Delta z \tag{17}$$

In fluid mechanics, the normal stress is related to the pressure and velocity gradient in the fluid, and is given as Eq. (18) [13]

$$\sigma_{xx} = -p + 2\mu\frac{\partial u}{\partial x} \tag{18}$$

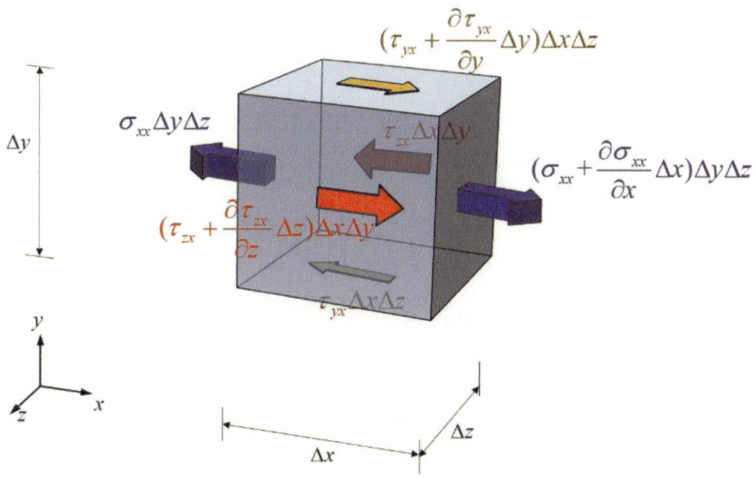

Figure 5.
The schematic of the surface forces along x direction on a control volume.

The shear stress follows Newton's law of viscosity as depicted in Eq. (16) [13]

$$\tau_{xy} = \tau_{yx} = \mu\left(\frac{\partial u}{\partial y} + \frac{\partial v}{\partial x}\right) \tag{19}$$

Bringing Eqs. (18) and (19) to Eq. (17) yields Eq. (20) for the total surface forces in the x direction.

$$\begin{aligned}\sum F_{\text{surf}} &= \frac{\partial}{\partial x}\left(-p + 2\mu\frac{\partial u}{\partial x}\right)\Delta x\Delta y\Delta z + \mu\frac{\partial}{\partial y}\left(\frac{\partial u}{\partial y} + \frac{\partial v}{\partial x}\right)\Delta x\Delta y\Delta z \\ &+\mu\frac{\partial}{\partial z}\left(\frac{\partial u}{\partial z} + \frac{\partial w}{\partial x}\right)\Delta x\Delta y\Delta z\end{aligned} \tag{20}$$

In most cases, the body force only includes gravity force.

$$\sum F_{\text{body}} = \rho\Delta x\Delta y\Delta z g_x \tag{21}$$

Then, the total force in the x direction is

$$\begin{aligned}\sum F_x &= \frac{\partial}{\partial x}\left(-p + 2\mu\frac{\partial u}{\partial x}\right)\Delta x\Delta y\Delta z + \mu\frac{\partial}{\partial y}\left(\frac{\partial u}{\partial y} + \frac{\partial v}{\partial x}\right)\Delta x\Delta y\Delta z \\ &+\mu\frac{\partial}{\partial z}\left(\frac{\partial u}{\partial z} + \frac{\partial w}{\partial x}\right)\Delta x\Delta y\Delta z + \rho\Delta x\Delta y\Delta z g_x\end{aligned} \tag{22}$$

Eqs. (22) and (16) yield

$$a_x = -\frac{1}{\rho}\frac{\partial p}{\partial x} + \frac{\mu}{\rho}\frac{\partial^2 u}{\partial x^2} + \frac{\mu}{\rho}\frac{\partial^2 u}{\partial y^2} + \frac{\mu}{\rho}\frac{\partial^2 u}{\partial z^2} + \frac{\mu}{\rho}\frac{\partial}{\partial x}\left(\frac{\partial u}{\partial x} + \frac{\partial v}{\partial y} + \frac{\partial w}{\partial z}\right) + g_x \tag{23}$$

The term inside the parenthesis is the expression of the continuity equation which equals zero

$$\frac{\partial u}{\partial x} + \frac{\partial v}{\partial y} + \frac{\partial w}{\partial z} = 0 \tag{24}$$

Then, Eq. (23) could be further simplified to

$$a_x = -\frac{1}{\rho}\frac{\partial p}{\partial x} + \frac{\mu}{\rho}\frac{\partial^2 u}{\partial x^2} + \frac{\mu}{\rho}\frac{\partial^2 u}{\partial y^2} + \frac{\mu}{\rho}\frac{\partial^2 u}{\partial z^2} + g_x \tag{25}$$

Combining Eqs. (15) and (25), we have

$$\frac{\partial u}{\partial t} + u\frac{\partial u}{\partial x} + v\frac{\partial u}{\partial y} + w\frac{\partial u}{\partial z} = -\frac{1}{\rho}\frac{\partial p}{\partial x} + \frac{\mu}{\rho}\left(\frac{\partial^2 u}{\partial x^2} + \frac{\partial^2 u}{\partial y^2} + \frac{\partial^2 u}{\partial z^2}\right) + g_x \tag{26}$$

Finally, Eq. (26) is the formula of the momentum equation of fluid in the x direction. Likewise, in the y direction and z direction, the momentum equations of fluid flow could also be derived in a similar way as follows:

$$\frac{\partial v}{\partial t}+u\frac{\partial v}{\partial x}+v\frac{\partial v}{\partial y}+w\frac{\partial v}{\partial z}=-\frac{1}{\rho}\frac{\partial p}{\partial y}+\frac{\mu}{\rho}\left(\frac{\partial^2 v}{\partial x^2}+\frac{\partial^2 v}{\partial y^2}+\frac{\partial^2 v}{\partial z^2}\right)+g_y \tag{27}$$

$$\frac{\partial w}{\partial t}+u\frac{\partial w}{\partial x}+v\frac{\partial w}{\partial y}+w\frac{\partial w}{\partial z}=-\frac{1}{\rho}\frac{\partial p}{\partial z}+\frac{\mu}{\rho}\left(\frac{\partial^2 w}{\partial x^2}+\frac{\partial^2 w}{\partial y^2}+\frac{\partial^2 w}{\partial z^2}\right)+g_z \tag{28}$$

Since Eq. (26) was progressively developed by Claude-Louis Navier and George Gabriel Stokes, this equation is generally called the Navier-Stokes Equation.

2.4 Energy equation

The heat transfer of fluid should follow the law of energy conservation. The total energy of fluid includes internal energy, kinetic energy and gravitational energy. Practically, it is not convenient to consider the total energy of fluid during the calculation. A better way to include kinetic energy and gravitational energy is by adding source terms in the energy equation. In a CFD model, the main energy form considered is thermal energy, so the thermal energy balance is usually considered for the energy equation. The thermal energy accumulated in an infinitesimal control volume equals the heat entering the control volume in terms of work, convection, conduction and radiation minus the heat going out. As a tradition, radiation is always not taken into account in the energy equation. The rate of energy change in a control volume should equal the net rate of heat added plus the net rate of work done. The heat added and work done could be either positive or negative depending on the direction of energy flow. The energy conservation is given as

$$\rho\frac{DE}{Dt}\Delta x\Delta y\Delta z=\sum\dot{Q}+\sum\dot{W} \tag{29}$$

where E is the energy contained in unit mass. For most fluid flow cases with heat exchange, reactions or adsorptions, temperature is the key parameter, so the energy of interest is usually thermal energy.

Since we have analyzed all the x-direction surface forces on a control volume when we derived the Navier-Stokes equation, the rate of heat generation due to work done by each normal force and shear force is shown in **Figure 6**. Likewise, there are also same number of normal forces and shear forces along y-direction and z-direction, and they generate heat as well.

Heat conduction caused by temperature gradient is also happening in heat transfer. The thermal energy entering and exiting the control volume in x-direction is described in **Figure 7**. Besides this pair of heat fluxes, in the pairs of x-o-z surfaces and x-o-y surfaces the conductive heat fluxes also exist.

Bring all the work heat and conductive heat into Eq. (29) and dividing $\Delta x\Delta y\Delta z$, we have

$$\begin{aligned}\rho\frac{DE}{Dt}=&-\frac{\partial q_x}{\partial x}-\frac{\partial q_y}{\partial y}-\frac{\partial q_z}{\partial z}+\frac{\partial(u\sigma_{xx})}{\partial x}+\frac{\partial(v\sigma_{yy})}{\partial y}\frac{\partial(w\sigma_{zz})}{\partial z}\\&+\frac{\partial(u\tau_{yx})}{\partial y}+\frac{\partial(u\tau_{zx})}{\partial z}+\frac{\partial(v\tau_{xy})}{\partial x}+\frac{\partial(v\tau_{zy})}{\partial z}+\frac{\partial(w\tau_{xz})}{\partial x}+\frac{\partial(w\tau_{yz})}{\partial y}\end{aligned} \tag{30}$$

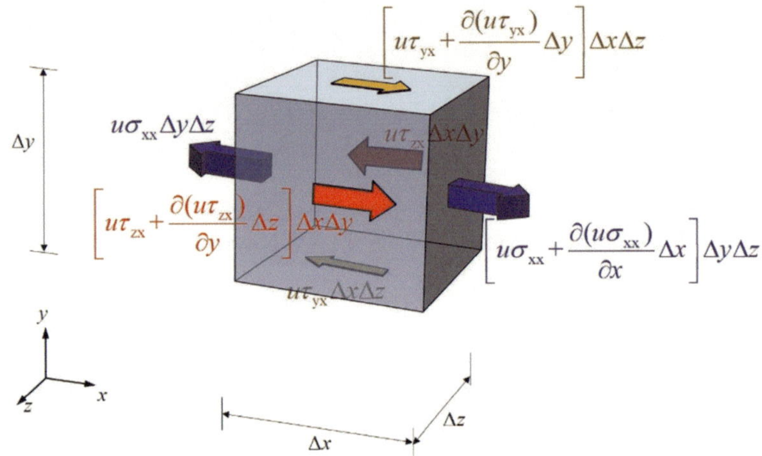

Figure 6.
The schematic of the work done by surface forces on a control volume.

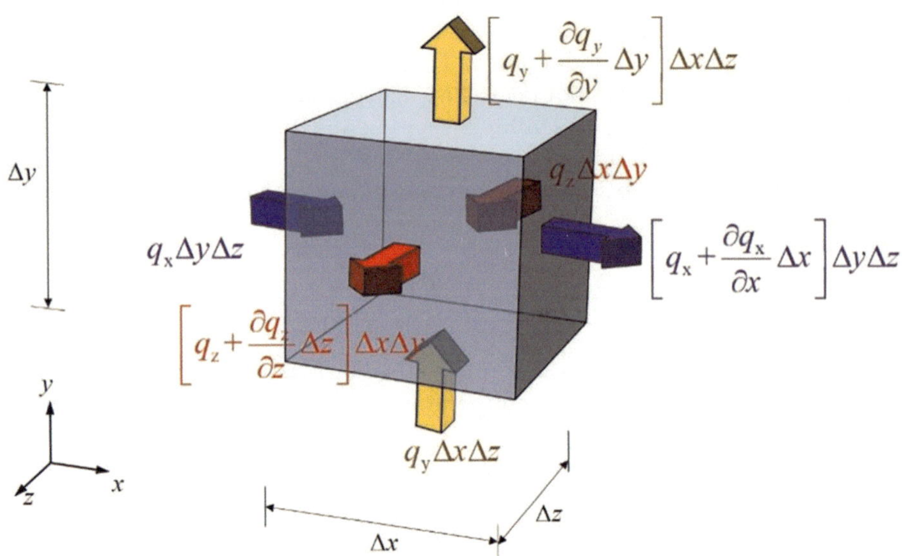

Figure 7.
The schematic of the conductive heat flow in a control volume.

The heat flux q normally follows Fourier's law of heat conduction which is given by [14]:

$$q_x = -\lambda \frac{\partial T}{\partial x} \tag{31}$$

$$q_y = -\lambda \frac{\partial T}{\partial y} \tag{32}$$

$$q_z = -\lambda \frac{\partial T}{\partial z} \tag{33}$$

Then, the energy balance equation (Eq. (30)) becomes Eq. (34)

$$\rho \frac{DE}{Dt} = \lambda \frac{\partial}{\partial x}\left(\frac{\partial T}{\partial x}\right) + \lambda \frac{\partial}{\partial y}\left(\frac{\partial T}{\partial y}\right) + \lambda \frac{\partial}{\partial z}\left(\frac{\partial T}{\partial z}\right) + \Phi \tag{34}$$

where Φ is the dissipation function which includes all terms relating to the energy due to deformation work done on the control volume. In most practical cases, dissipation function Φ can be neglected, because the term is significantly small compared with heat conduction. Then, Eq. (34) is simplified as

$$\rho \frac{DE}{Dt} = \lambda \frac{\partial}{\partial x}\left(\frac{\partial T}{\partial x}\right) + \lambda \frac{\partial}{\partial y}\left(\frac{\partial T}{\partial y}\right) + \lambda \frac{\partial}{\partial z}\left(\frac{\partial T}{\partial z}\right) \tag{35}$$

If the energy E only refers to internal energy, the change of energy DE could be represented by C_pDT. Then, Eq. (35) could be further rearranged to Eq. (36)

$$\rho C_p \frac{DT}{Dt} = \lambda \frac{\partial}{\partial x}\left(\frac{\partial T}{\partial x}\right) + \lambda \frac{\partial}{\partial y}\left(\frac{\partial T}{\partial y}\right) + \lambda \frac{\partial}{\partial z}\left(\frac{\partial T}{\partial z}\right) \tag{36}$$

Expanding the total derivative of T, we get

$$\frac{\partial T}{\partial t} + u\frac{\partial T}{\partial x} + v\frac{\partial T}{\partial y} + w\frac{\partial T}{\partial z} = \frac{\lambda}{\rho C_p}\left(\frac{\partial^2 T}{\partial x^2} + \frac{\partial^2 T}{\partial y^2} + \frac{\partial^2 T}{\partial z^2}\right) \tag{37}$$

Eq. (37) is the final expression of the energy equation. The first term on the left-hand side is called the transient term, the second to the fourth terms are called advection terms, and the terms on the right-hand sides are diffusion terms. Energy equation related temperature to time and space coordinate. With the energy equation, obtaining the temperature evolution and distribution in a fluid field becomes possible.

2.5 Summary of the governing equations

Previously, we have derived the continuity equation, component equation, momentum equation and energy equation based on the conservation principles.

$$\frac{\partial \rho}{\partial t} + \frac{\partial(\rho u)}{\partial x} + \frac{\partial(\rho v)}{\partial y} + \frac{\partial(\rho w)}{\partial z} = 0 \tag{38}$$

$$\frac{\partial(\rho X_i)}{\partial t} + \frac{\partial(\rho u X_i)}{\partial x} + \frac{\partial(\rho v X_i)}{\partial y} + \frac{\partial(\rho w X_i)}{\partial z} = D\left[\frac{\partial^2(\rho X_i)}{\partial x^2} + \frac{\partial^2(\rho X_i)}{\partial y^2} + \frac{\partial^2(\rho X_i)}{\partial z^2}\right] \tag{39}$$

$$\frac{\partial(\rho u)}{\partial t} + \frac{\partial(\rho uu)}{\partial x} + \frac{\partial(\rho vu)}{\partial y} + \frac{\partial(\rho wu)}{\partial z} = -\frac{\partial p}{\partial x} + \mu\left(\frac{\partial^2 u}{\partial x^2} + \frac{\partial^2 u}{\partial y^2} + \frac{\partial^2 u}{\partial z^2}\right) + g_x \tag{40}$$

$$\frac{\partial(\rho v)}{\partial t} + \frac{\partial(\rho uv)}{\partial x} + \frac{\partial(\rho vv)}{\partial y} + \frac{\partial(\rho wv)}{\partial z} = -\frac{\partial p}{\partial y} + \mu\left(\frac{\partial^2 v}{\partial x^2} + \frac{\partial^2 v}{\partial y^2} + \frac{\partial^2 v}{\partial z^2}\right) + g_y \tag{41}$$

$$\frac{\partial(\rho w)}{\partial t} + \frac{\partial(\rho uw)}{\partial x} + \frac{\partial(\rho vw)}{\partial y} + \frac{\partial(\rho ww)}{\partial z} = -\frac{\partial p}{\partial z} + \mu\left(\frac{\partial^2 w}{\partial x^2} + \frac{\partial^2 w}{\partial y^2} + \frac{\partial^2 w}{\partial z^2}\right) + g_z \tag{42}$$

$$\frac{\partial(\rho T)}{\partial t} + \frac{\partial(\rho uT)}{\partial x} + \frac{\partial(\rho vT)}{\partial y} + \frac{\partial(\rho wT)}{\partial z} = \frac{\lambda}{C_p}\left(\frac{\partial^2 T}{\partial x^2} + \frac{\partial^2 T}{\partial y^2} + \frac{\partial^2 T}{\partial z^2}\right) \tag{43}$$

If we introduce a universal variable ϕ, a generic form of the governing equations could be written as follows,

$$\frac{\partial(\rho\phi)}{\partial t}+\frac{\partial(\rho u\phi)}{\partial x}+\frac{\partial(\rho v\phi)}{\partial y}+\frac{\partial(\rho w\phi)}{\partial z}=\Gamma\left(\frac{\partial^2\phi}{\partial x^2}+\frac{\partial^2\phi}{\partial y^2}+\frac{\partial^2\phi}{\partial z^2}\right) \tag{44}$$

where Γ is the diffusion coefficient. If $\phi = 1$, Eq. (44) is the continuity equation. If $\phi = X_i$, and $\Gamma = D$, Eq. (44) transforms to the component equation. If $\phi = u$, v or w, $\Gamma = \mu$, and the pressure term and gravity term are represented by source terms, Eq. (44) becomes the Navier-Stokes equation. If $\phi = T$, and $\Gamma = \lambda/C_p$, Eq. (44) turns out to be the energy equation.

2.6 Governing equations in cylindrical coordinate

In some common cases, fluid flows through pipes or reactors which are mostly in cylindrical shape, so it is more convenient to express the flow in a cylindrical coordinate. Writing CFD code based on the governing equations expressed in cylindrical coordinate is also more suitable than those in Cartesian coordinate [10].

The continuity equation could be derived from analyzing the mass balance in the fluid microelement as shown in **Figure 8**. The blue arrows represent the flow in radial direction. The red color arrows represent the flow in the tangential direction, and the yellow arrows stand for the flow in the axial direction. Therefore, the conservation of mass in the control volume in **Figure 8** is given as

$$\begin{aligned}\frac{\partial(\rho r\Delta r\Delta\theta\Delta z)}{\partial t} &= u_r(r\Delta\theta\Delta z)-\left(u_r+\frac{\partial u_r}{\partial r}\Delta r\right)(r+\Delta r)\Delta\theta\Delta z\\ &+u_\theta(\Delta r\Delta z)-\left(u_\theta+\frac{\partial u_\theta}{\partial\theta}\Delta\theta\right)(\Delta r\Delta z)\\ &+u_z\left[\left(r+\frac{\Delta r}{2}\right)\Delta r\Delta\theta\right]-\left(u_z+\frac{\partial u_z}{\partial z}\Delta z\right)\left[\left(r+\frac{\Delta r}{2}\right)\Delta r\Delta\theta\right]\end{aligned} \tag{45}$$

Dividing $r\Delta r\Delta\theta\Delta z$ on both sides, and taking the limit for $\Delta r \to 0$, $\Delta\theta \to 0$ and $\Delta z \to 0$, Eq. (45) can be simplified to Eq. (46), which is the continuity equation in cylindrical coordinate.

$$\frac{\partial\rho}{\partial t}+\frac{1}{r}\frac{\partial(ru_r)}{\partial r}+\frac{1}{r}\frac{\partial u_\theta}{\partial\theta}+\frac{\partial u_z}{\partial z}=0 \tag{46}$$

The momentum equation in cylindrical coordinate is also derived from Newton's second law.

$$\rho\vec{\mathbf{a}}=\frac{\vec{\mathbf{F}}}{V}=\frac{\overrightarrow{\mathbf{F_{surf}}}+\overrightarrow{\mathbf{F_{body}}}}{V}=\frac{\overrightarrow{\mathbf{F_{surf}}}}{V}+\rho\vec{\mathbf{g}} \tag{47}$$

which is then expressed by velocity vector as

$$\frac{\mathrm{D}\vec{\mathbf{u}}}{\mathrm{D}t}=\frac{1}{\rho}\frac{\overrightarrow{\mathbf{F_{surf}}}}{V}+\vec{\mathbf{g}} \tag{48}$$

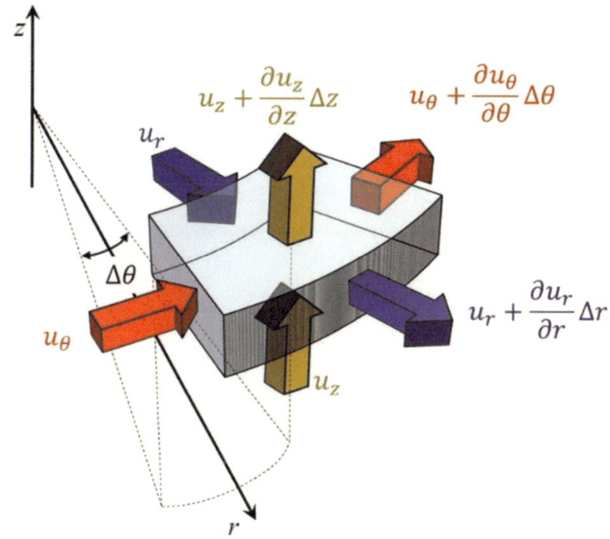

Figure 8.
The schematic of mass flux of a control volume in cylindrical coordinate.

In cylindrical coordinate, the velocity vector is given as

$$\vec{\mathbf{u}}(r,\theta,z,t) = u_r(r,\theta,z,t)\overrightarrow{\mathbf{e}_r} + u_\theta(r,\theta,z,t)\overrightarrow{\mathbf{e}_\theta} + u_z(r,\theta,z,t)\overrightarrow{\mathbf{e}_z} \tag{49}$$

And the acceleration is then given by

$$\vec{\mathbf{a}} = \frac{\mathrm{D}\vec{\mathbf{u}}}{\mathrm{D}t} = \frac{\partial\vec{\mathbf{u}}}{\partial t} + \frac{\partial\vec{\mathbf{u}}}{\partial r}\frac{\mathrm{d}r}{\mathrm{d}t} + \frac{\partial\vec{\mathbf{u}}}{\partial\theta}\frac{\mathrm{d}\theta}{\mathrm{d}t} + \frac{\partial\vec{\mathbf{u}}}{\partial z}\frac{\mathrm{d}z}{\mathrm{d}t} \tag{50}$$

The partial derivative of velocity vector with respect to time can be expressed as the summation of three vectors in radial ®, tangential (θ) and axial (z) directions (**Figure 9**) and the definition of u_r, u_θ and u_z is.

$$\frac{\partial\vec{\mathbf{u}}}{\partial t} = \frac{\partial u_r}{\partial t}\overrightarrow{\mathbf{e}_r} + \frac{\partial u_\theta}{\partial t}\overrightarrow{\mathbf{e}_\theta} + \frac{\partial u_z}{\partial t}\overrightarrow{\mathbf{e}_z} \tag{51}$$

$$u_r = \frac{\mathrm{d}r}{\mathrm{d}t} \tag{52}$$

$$u_\theta = \frac{r\mathrm{d}\theta}{\mathrm{d}t} \tag{53}$$

$$u_z = \frac{\mathrm{d}z}{\mathrm{d}t} \tag{54}$$

With the definitions in Eq. (52), Eq. (50) is then transformed to

$$\frac{\mathrm{D}\vec{\mathbf{u}}}{\mathrm{D}t} = \frac{\partial\vec{\mathbf{u}}}{\partial t} + u_r\frac{\partial\vec{\mathbf{u}}}{\partial r} + \frac{u_\theta}{r}\frac{\partial\vec{\mathbf{u}}}{\partial\theta} + u_z\frac{\partial\vec{\mathbf{u}}}{\partial z} \tag{55}$$

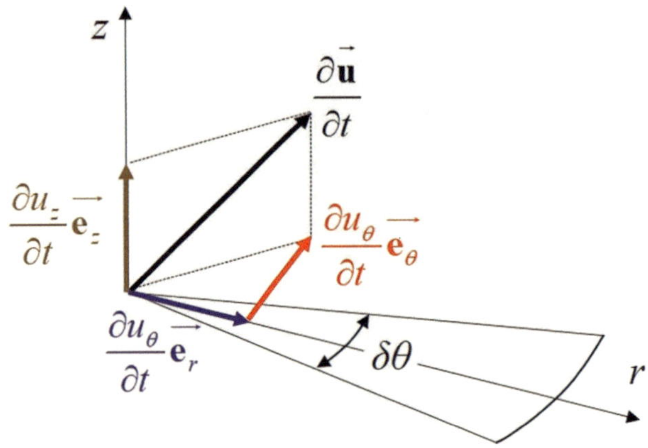

Figure 9.
The schematic of $\partial\mathbf{u}/\partial t$ *in cylindrical coordinate.*

The partial derivatives of velocity vector with respect to r, θ and z are

$$\frac{\partial \vec{\mathbf{u}}}{\partial r} = \frac{\partial\left(u_r\vec{\mathbf{e}_r} + u_\theta\vec{\mathbf{e}_\theta} + u_z\vec{\mathbf{e}_z}\right)}{\partial r} = \frac{\partial u_r}{\partial r}\vec{\mathbf{e}_r} + u_r\frac{\partial\vec{\mathbf{e}_r}}{\partial r} + \frac{\partial u_\theta}{\partial r}\vec{\mathbf{e}_\theta} + u_\theta\frac{\partial\vec{\mathbf{e}_\theta}}{\partial r} + \frac{\partial u_z}{\partial r}\vec{\mathbf{e}_z} + u_z\frac{\partial\vec{\mathbf{e}_z}}{\partial r} \tag{56}$$

$$\frac{\partial \vec{\mathbf{u}}}{\partial \theta} = \frac{\partial\left(u_r\vec{\mathbf{e}_r} + u_\theta\vec{\mathbf{e}_\theta} + u_z\vec{\mathbf{e}_z}\right)}{\partial \theta} = \frac{\partial u_r}{\partial \theta}\vec{\mathbf{e}_r} + u_r\frac{\partial\vec{\mathbf{e}_r}}{\partial \theta} + \frac{\partial u_\theta}{\partial \theta}\vec{\mathbf{e}_\theta} + u_\theta\frac{\partial\vec{\mathbf{e}_\theta}}{\partial \theta} + \frac{\partial u_z}{\partial \theta}\vec{\mathbf{e}_z} + u_z\frac{\partial\vec{\mathbf{e}_z}}{\partial \theta} \tag{57}$$

$$\frac{\partial \vec{\mathbf{u}}}{\partial z} = \frac{\partial\left(u_r\vec{\mathbf{e}_r} + u_\theta\vec{\mathbf{e}_\theta} + u_z\vec{\mathbf{e}_z}\right)}{\partial z} = \frac{\partial u_r}{\partial z}\vec{\mathbf{e}_r} + u_r\frac{\partial\vec{\mathbf{e}_r}}{\partial z} + \frac{\partial u_\theta}{\partial z}\vec{\mathbf{e}_\theta} + u_\theta\frac{\partial\vec{\mathbf{e}_\theta}}{\partial z} + \frac{\partial u_z}{\partial z}\vec{\mathbf{e}_z} + u_z\frac{\partial\vec{\mathbf{e}_z}}{\partial z} \tag{58}$$

Particularly, in cylindrical coordinate the unit vectors in radial (r), tangential (θ) and axial (z) directions have the following correlations

$$\frac{\partial\vec{\mathbf{e}_r}}{\partial r} = \frac{\partial\vec{\mathbf{e}_\theta}}{\partial r} = \frac{\partial\vec{\mathbf{e}_z}}{\partial r} = 0 \tag{59}$$

$$\frac{\partial\vec{\mathbf{e}_r}}{\partial \theta} = \vec{\mathbf{e}_\theta} \tag{60}$$

$$\frac{\partial\vec{\mathbf{e}_\theta}}{\partial \theta} = -\vec{\mathbf{e}_r} \tag{61}$$

Employing the correlations in Eq. (59), Eq. (56) would become

$$\frac{\partial \vec{\mathbf{u}}}{\partial r} = \frac{\partial u_r}{\partial r}\vec{\mathbf{e}_r} + \frac{\partial u_\theta}{\partial r}\vec{\mathbf{e}_\theta} + \frac{\partial u_z}{\partial r}\vec{\mathbf{e}_z} \tag{62}$$

$$\frac{\partial \vec{\mathbf{u}}}{\partial \theta} = \left(\frac{\partial u_r}{\partial \theta} - u_\theta\right)\vec{\mathbf{e}_r} + \left(u_r + \frac{\partial u_\theta}{\partial \theta}\right)\vec{\mathbf{e}_\theta} + \frac{\partial u_z}{\partial \theta}\vec{\mathbf{e}_z} \tag{63}$$

$$\frac{\partial \vec{\mathbf{u}}}{\partial z} = \frac{\partial u_r}{\partial z}\vec{\mathbf{e}_r} + \frac{\partial u_\theta}{\partial z}\vec{\mathbf{e}_\theta} + \frac{\partial u_z}{\partial z}\vec{\mathbf{e}_z} \tag{64}$$

Bringing Eqs. (51) and (62) to Eq. (55), we can obtain

$$\begin{aligned}
\frac{\mathrm{D}\vec{\mathbf{u}}}{\mathrm{D}t} &= \frac{\partial \vec{\mathbf{u}}}{\partial t} + \frac{\partial \vec{\mathbf{u}}}{\partial r}\frac{\mathrm{d}r}{\mathrm{d}t} + \frac{\partial \vec{\mathbf{u}}}{\partial \theta}\frac{\mathrm{d}\theta}{\mathrm{d}t} + \frac{\partial \vec{\mathbf{u}}}{\partial z}\frac{\mathrm{d}z}{\mathrm{d}t} \\
&= \frac{\partial u_r}{\partial t}\vec{\mathbf{e}_r} + \frac{\partial u_\theta}{\partial t}\vec{\mathbf{e}_\theta} + \frac{\partial u_z}{\partial t}\vec{\mathbf{e}_z} \\
&+ u_r\left[\frac{\partial u_r}{\partial r}\vec{\mathbf{e}_r} + \frac{\partial u_\theta}{\partial r}\vec{\mathbf{e}_\theta} + \frac{\partial u_z}{\partial r}\vec{\mathbf{e}_z}\right] \\
&+ \frac{u_\theta}{r}\left[\left(\frac{\partial u_r}{\partial \theta} - u_\theta\right)\vec{\mathbf{e}_r} + \left(u_r + \frac{\partial u_\theta}{\partial \theta}\right)\vec{\mathbf{e}_\theta} + \frac{\partial u_z}{\partial \theta}\vec{\mathbf{e}_z}\right] \\
&+ u_z\left[\frac{\partial u_r}{\partial z}\vec{\mathbf{e}_r} + \frac{\partial u_\theta}{\partial z}\vec{\mathbf{e}_\theta} + \frac{\partial u_z}{\partial z}\vec{\mathbf{e}_z}\right] \\
&= \left(\frac{\partial u_r}{\partial t} + u_r\frac{\partial u_r}{\partial r} + \frac{u_\theta}{r}\frac{\partial u_r}{\partial \theta} - \frac{u_\theta^2}{r} + u_z\frac{\partial u_r}{\partial z}\right)\vec{\mathbf{e}_r} \\
&+ \left(\frac{\partial u_\theta}{\partial t} + u_r\frac{\partial u_\theta}{\partial r} + \frac{u_\theta}{r}\frac{\partial u_\theta}{\partial \theta} + \frac{u_\theta u_r}{r} + u_z\frac{\partial u_\theta}{\partial z}\right)\vec{\mathbf{e}_\theta} \\
&+ \left(\frac{\partial u_z}{\partial t} + u_r\frac{\partial u_z}{\partial r} + \frac{u_\theta}{r}\frac{\partial u_z}{\partial \theta} + u_z\frac{\partial u_z}{\partial z}\right)\vec{\mathbf{e}_z}
\end{aligned} \tag{65}$$

Now, we took the radial direction as an example to derive its momentum equation. The acceleration in the radial direction is defined as

$$a_r = \frac{\mathrm{D}u_r}{\mathrm{D}t} = \frac{\partial u_r}{\partial t} + u_r\frac{\partial u_r}{\partial r} + \frac{u_\theta}{r}\frac{\partial u_r}{\partial \theta} - \frac{u_\theta^2}{r} \tag{66}$$

According to Newton's law, the acceleration also equals to

$$a_r = \frac{1}{\rho}\frac{F_r}{\delta V} + g_r \tag{67}$$

The surface tensions in the radial direction are described in **Fig. 10**. The summation of all the surface tensions along the radial direction is

$$\begin{aligned}
F_r &= \sigma_{rr}(r+\Delta r)\Delta\theta\Delta z + \frac{\partial \sigma_{rr}}{\partial r}\Delta r(r+\Delta r)\Delta\theta\Delta z - \sigma_{rr}(r\Delta\theta\Delta z) \\
&+ \sigma_{\theta r}(\Delta r\Delta z)\cos\frac{\Delta\theta}{2} + \frac{\partial \sigma_{\theta r}}{\partial \theta}\Delta\theta(\Delta r\Delta z)\cos\frac{\Delta\theta}{2} - \sigma_{\theta r}(\Delta r\Delta z)\cos\frac{\Delta\theta}{2} \\
&+ \sigma_{zr}\left[\left(r+\frac{\Delta r}{2}\right)\Delta r\Delta\theta\right] + \frac{\partial \sigma_{zr}}{\partial z}\Delta z\left[\left(r+\frac{\Delta r}{2}\right)\Delta r\Delta\theta\right] - \sigma_{rzrz}\left[\left(r+\frac{\Delta r}{2}\right)\Delta r\Delta\theta\right] \\
&- \sigma_{\theta\theta}(\Delta r\Delta z)\sin\frac{\Delta\theta}{2} - \frac{\partial \sigma_{\theta\theta}}{\partial \theta}\Delta\theta(\Delta r\Delta z)\sin\frac{\Delta\theta}{2} - \sigma_{\theta\theta}(\Delta r\Delta z)\sin\frac{\Delta\theta}{2}
\end{aligned} \tag{68}$$

Taking the limit for $\Delta r \to 0$, $\Delta\theta \to 0$ and $\Delta z \to 0$, Eq. (68) is then achieved to Eq. (69) which is the continuity equation in cylindrical coordinate.

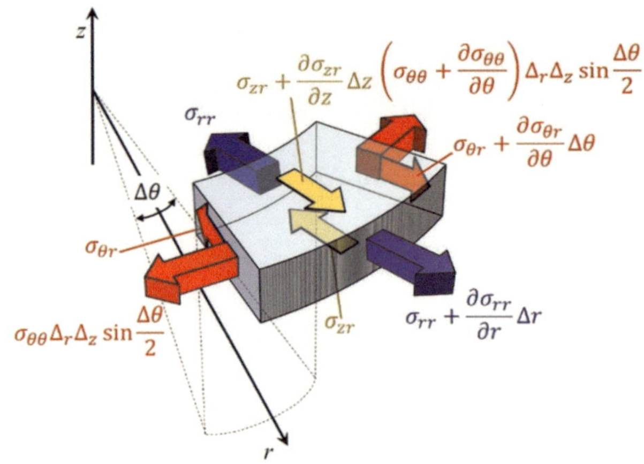

Figure 10.
The schematic of the surface tensions along radial direction on a control volume in cylindrical coordinate.

$$
\begin{aligned}
F_r = {} & \sigma_{rr}\Delta r \Delta\theta \Delta z + \frac{\partial \sigma_{rr}}{\partial r}\Delta r (r + \Delta r)\Delta\theta \Delta z \\
& + \frac{\partial \sigma_{\theta r}}{\partial \theta}\Delta\theta (\Delta r \Delta z)\cos\frac{\Delta\theta}{2} \\
& + \frac{\partial \sigma_{zr}}{\partial z}\Delta z\left[\left(r + \frac{\Delta r}{2}\right)\Delta r \Delta\theta\right] \\
& - 2\sigma_{\theta\theta}(\Delta r \Delta z)\frac{\Delta\theta}{2} - \frac{\partial \sigma_{\theta\theta}}{\partial \theta}\Delta\theta (\Delta r \Delta z)\frac{\Delta\theta}{2}
\end{aligned}
\tag{69}
$$

Bringing $\Delta V = r\Delta r \Delta\theta \Delta z$ into Eq. (69) gives

$$
\frac{F_r}{\Delta V} = \frac{\sigma_{rr}}{r} + \frac{\partial \sigma_{rr}}{\partial r} + \frac{1}{r}\frac{\partial \sigma_{\theta r}}{\partial \theta} + \frac{\partial \sigma_{zr}}{\partial z} - \frac{\sigma_{\theta\theta}}{r} \tag{70}
$$

Combining Eqs. (66), (67) and (70), we can obtain

$$
\frac{\partial u_r}{\partial t} + u_r \frac{\partial u_r}{\partial r} + \frac{u_\theta}{r}\frac{\partial u_r}{\partial \theta} - \frac{u_\theta^2}{r} + u_z \frac{\partial u_r}{\partial z} = \frac{1}{\rho}\left(\frac{\sigma_{rr}}{r} + \frac{\partial \sigma_{rr}}{\partial r} + \frac{1}{r}\frac{\partial \sigma_{\theta r}}{\partial \theta} + \frac{\partial \sigma_{zr}}{\partial z} - \frac{\sigma_{\theta\theta}}{r}\right) + g_r \tag{71}
$$

In fluid mechanics, the definitions of the surface tensions in Eq. (70) are given as follows:

$$
\sigma_{rr} = -p + 2\mu \frac{\partial u_r}{\partial r} \tag{72}
$$

$$
\sigma_{\theta\theta} = -p + 2\mu\left(\frac{1}{r}\frac{\partial u_\theta}{\partial \theta} + \frac{u_r}{r}\right) \tag{73}
$$

$$
\sigma_{zz} = -p + 2\mu \frac{\partial u_z}{\partial z} \tag{74}
$$

$$
\sigma_{r\theta} = \sigma_{\theta r} = \mu\left(\frac{1}{r}\frac{\partial u_r}{\partial \theta} + \frac{\partial u_\theta}{\partial r} - \frac{u_\theta}{r}\right) \tag{75}
$$

$$\sigma_{rz} = \sigma_{zr} = \mu\left(\frac{\partial u_r}{\partial z} + \frac{\partial u_z}{\partial r}\right) \tag{76}$$

$$\sigma_{\theta z} = \sigma_{z\theta} = \mu\left(\frac{1}{r}\frac{\partial u_z}{\partial \theta} + \frac{\partial u_\theta}{\partial z}\right) \tag{77}$$

Combining Eqs. (70) and (71) yields

$$\begin{aligned}&\frac{\partial u_r}{\partial t} + u_r\frac{\partial u_r}{\partial r} + \frac{u_\theta}{r}\frac{\partial u_r}{\partial \theta} - \frac{u_\theta^2}{r} + u_z\frac{\partial u_r}{\partial z}\\ &= -\frac{1}{\rho}\frac{\partial p}{\partial r} + \frac{\mu}{\rho}\left[-\frac{u_r}{r^2} + \frac{1}{r}\frac{\partial}{\partial r}\left(r\frac{\partial u_r}{\partial r}\right) + \frac{1}{r^2}\frac{\partial^2 u_r}{\partial \theta^2} + \frac{\partial^2 u_r}{\partial z^2} - \frac{2}{r^2}\frac{\partial u_\theta}{\partial \theta}\right] + g_r\end{aligned} \tag{78}$$

Finally, Eq. (78) is the momentum equation of fluid in the radial direction. Likewise, in the tangential and axial directions, the momentum equations of fluid flow could also be derived in a similar way as follows:

$$\begin{aligned}&\frac{\partial u_\theta}{\partial t} + u_r\frac{\partial u_\theta}{\partial r} + \frac{u_\theta}{r}\frac{\partial u_\theta}{\partial \theta} - \frac{u_r u_\theta}{r} + u_z\frac{\partial u_\theta}{\partial z}\\ &= -\frac{1}{\rho}\frac{\partial p}{\partial \theta} + \frac{\mu}{\rho}\left[-\frac{u_\theta}{r^2} + \frac{1}{r}\frac{\partial}{\partial r}\left(r\frac{\partial u_\theta}{\partial r}\right) + \frac{1}{r^2}\frac{\partial^2 u_\theta}{\partial \theta^2} + \frac{\partial^2 u_\theta}{\partial z^2} + \frac{2}{r^2}\frac{\partial u_r}{\partial \theta}\right] + g_\theta\end{aligned} \tag{79}$$

$$\begin{aligned}&\frac{\partial u_z}{\partial t} + u_r\frac{\partial u_z}{\partial r} + \frac{u_\theta}{r}\frac{\partial u_z}{\partial \theta} + u_z\frac{\partial u_z}{\partial z}\\ &= -\frac{1}{\rho}\frac{\partial p}{\partial z} + \frac{\mu}{\rho}\left[\frac{1}{r}\frac{\partial}{\partial r}\left(r\frac{\partial u_z}{\partial r}\right) + \frac{1}{r^2}\frac{\partial^2 u_z}{\partial \theta^2} + \frac{\partial^2 u_z}{\partial z^2}\right] + g_z\end{aligned} \tag{80}$$

The energy equation could be derived from analyzing the energy balance in the fluid control volume as shown in **Figure 11**.

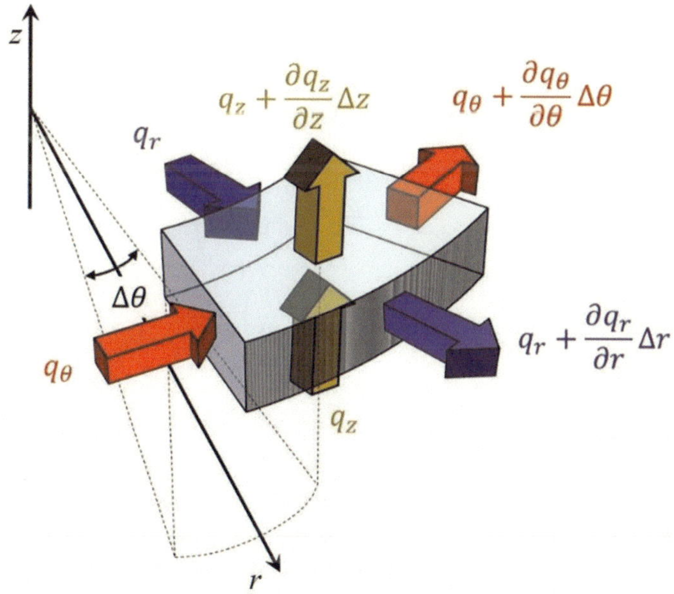

Figure 11.
The schematic of the conductive heat fluxes in a control volume in cylindrical coordinate.

According to **Figure 11** for cylindrical coordinate, the heat energy balance is expressed as the energy accumulated in the control volume equals the heat into this control volume minus the heat out of the control volume

$$\begin{aligned}\rho\frac{DE}{Dt}r\Delta r\Delta\theta\Delta z =& -\left(q_r+\frac{\partial q_r}{\partial r}\Delta r\right)(r+\Delta r)\Delta\theta\Delta z + q_r(r\Delta\theta\Delta z)\\ &-\left(q_\theta+\frac{\partial q_\theta}{\partial\theta}\Delta\theta\right)(\Delta r\Delta z)\cos\frac{\Delta\theta}{2}+q_\theta(\Delta r\Delta z)\cos\frac{\Delta\theta}{2}\\ &-\left(q_z+\frac{\partial q_z}{\partial z}\Delta z\right)\left[\left(r+\frac{\Delta r}{2}\right)\Delta r\Delta\theta\right]+q_z\left[\left(r+\frac{\Delta r}{2}\right)\Delta r\Delta\theta\right]\end{aligned} \tag{81}$$

Which can be further transformed to Eq. (82) by canceling out identical terms with opposite signs.

$$\rho\frac{DE}{Dt}=-\frac{1}{r}\frac{\partial(rq_r)}{\partial r}-\frac{1}{r}\frac{\partial q_\theta}{\partial\theta}-\frac{\partial q_z}{\partial z} \tag{82}$$

If the change of thermal energy is all induced by temperature change, then Eq. (82) is expressed as:

$$\rho C_\mathrm{p}\left(\frac{\partial T}{\partial t}+u_r\frac{\partial T}{\partial r}+\frac{u_\theta}{r}\frac{\partial T}{\partial\theta}+u_z\frac{\partial T}{\partial z}\right)=-\frac{1}{r}\frac{\partial(rq_r)}{\partial r}-\frac{1}{r}\frac{\partial q_\theta}{\partial\theta}-\frac{\partial q_z}{\partial z} \tag{83}$$

By applying Fourier's law of heat conduction (Eq. 84), Eq. (89) is then derived as the energy equation in cylindrical coordinate.

$$q_r = -\lambda\frac{\partial T}{\partial r} \tag{84}$$

$$q_\theta = -\frac{\lambda}{r}\frac{\partial T}{\partial\theta} \tag{85}$$

$$q_z = -\lambda\frac{\partial T}{\partial z} \tag{86}$$

$$\begin{aligned}&\rho C_\mathrm{p}\left(\frac{\partial T}{\partial t}+u_r\frac{\partial T}{\partial r}+\frac{u_\theta}{r}\frac{\partial T}{\partial\theta}+u_z\frac{\partial T}{\partial z}\right)\\ &=\lambda\left[\frac{1}{r}\frac{\partial}{\partial r}\left(r\frac{\partial T}{\partial r}\right)+\frac{1}{r^2}\frac{\partial}{\partial\theta}\left(\frac{\partial T}{\partial\theta}\right)+\frac{\partial}{\partial z}\left(\frac{\partial T}{\partial z}\right)\right]\end{aligned} \tag{87}$$

3. Source and sinks

The expansion of CFD applications to the fields such as chemical engineering, environmental engineering, thermal engineering and pharmaceutical engineering was based on the development of multiscale and multiphase models which benefited from the coupling of source or sink terms to the conservation equations. A sink term is actually a negative source term, so both source terms and sink terms could be called by a joint name—source terms.

3.1 Source term for the continuity equation

Processes such as solid gasification or combustion, and gas adsorption by solid sorbents would break the mass conservation of the fluid phase. If there are reactions are generating additional fluid or remove the existing fluid in a control volume, the mass balance in this control volume must be disrupted by these reactions. The rate of fluid entering a control volume minus the rate of fluid flowing out equals the rate of fluid accumulating in this control volume and the net gain of fluid generated inside the control volume. If in the control volume there is a reaction generating fluid at a rate of r, Eq. (1) for mass balance will be altered to

$$\frac{dm}{dt} = r dx dy dz + \sum_{in} \dot{m} - \sum_{out} \dot{m} \tag{88}$$

Eq. (45) could also include the case where there is a depletion of fluid by using a negative value for r. Therefore, the continuity equation is adjusted by a source term as follows.

$$\frac{\partial \rho}{\partial t} + \frac{\partial(\rho u)}{\partial x} + \frac{\partial(\rho v)}{\partial y} + \frac{\partial(\rho w)}{\partial z} = S_{continuity} \tag{89}$$

where the source term is the reaction rate $S_{continuity} = r$ (kg m^{-3} s^{-1}) if the reaction is a body reaction. For a surface reaction, the source term is the multiplication of the surface reaction rate (kg m^{-2} s^{-1}) with the reaction area in unit volume (A_{react}/V_{body}, m^2 m^{-3}) [10].

3.2 Source term for component equation

Similar to the imbalance for total mass conservation, the conservation of each component in a multi-component flow with reactions could also be broken. The components participating as reactants would be deemed as sinks and the components generated as products would be regarded as sources. If a component is generated at a rate of r_i, then Eq. (7) is adjusted accordingly as follows

$$\frac{dm_i}{dt} = r_i dx dy dz + \sum_{in} \dot{m}_i - \sum_{out} \dot{m}_i \tag{90}$$

After a similar derivation from Eq. (7) to Eq. (9), the component equation should include a source term as expressed by Eq. (91)

$$\begin{aligned} &\frac{\partial(\rho X_i)}{\partial t} + \frac{\partial(\rho u X_i)}{\partial x} + \frac{\partial(\rho v X_i)}{\partial y} + \frac{\partial(\rho w X_i)}{\partial z} \\ &= D\left[\frac{\partial^2(\rho X_i)}{\partial x^2} + \frac{\partial^2(\rho X_i)}{\partial y^2} + \frac{\partial^2(\rho X_i)}{\partial z^2}\right] + S_i \end{aligned} \tag{91}$$

where S_i is the source term of the component equation, and it equals reaction rate r_i if it is a body reaction. If it is a surface reaction, the source term of Eq. (91) equals the multiplication of the surface reaction rate (kg m^{-2} s^{-1}) with the reaction area in unit volume (A_{react}/V_{body}, m^2 m^{-3}).

3.3 Source term for the momentum equation

When it comes to cases with a multiphase flow such as gas-solid flow in fluidized bed, momentum conservation still holds, but interphase interactions need to be taken into account [15]. There might be extra forces between different phases, which can affect the momentum transfer between different phases. Therefore, in multiphase flow, appropriate interphase force terms need to be introduced to correct the momentum equation.

When the fluid phase and solid phase interact with each other, a reactive force that causes flow resistance is generated, known as drag force. If the drag force is expressed as $F_{\mathrm{D,i}}$, Eq. (13) can be expanded as follows

$$\frac{\mathrm{D}u}{\mathrm{D}t}=\frac{\sum F_x}{m}=\frac{\sum F_{\mathrm{surf}}+\sum F_{\mathrm{body}}+\sum F_{\mathrm{D},i}}{m} \tag{92}$$

The drag force $\boldsymbol{F}_{\mathrm{D,I}}$ is calculated by:

$$\boldsymbol{F}_{\mathrm{D},i}=\frac{\beta_{\mathrm{fs}}}{\theta_{\mathrm{f}}}(\boldsymbol{u}_{\mathrm{s}}-\boldsymbol{u}_{\mathrm{f}}) \tag{93}$$

where $\boldsymbol{u}_{\mathrm{s}}$ is the solid phase velocity, $\boldsymbol{u}_{\mathrm{f}}$ is the fluid phase velocity, θ_{f} is the volume fraction of the fluid phase and β_{fs} is the drag coefficient. Many researchers have proposed the models for drag coefficients. The most widely used is the Gidaspow model which combines Ergun and Wen-Yu equations to accurately simulate the gas-solid multiphase flow:

$$\beta_{\mathrm{p,f}}=\begin{cases}150\dfrac{\theta_{\mathrm{s}}^2\mu_{\mathrm{f}}}{\theta_{\mathrm{f}}^2 d_{\mathrm{s}}}+1.75\dfrac{\theta_{\mathrm{s}}\rho_{\mathrm{f}}}{\theta_{\mathrm{f}} d_{\mathrm{s}}}|\boldsymbol{u}_{\mathrm{s}}-\boldsymbol{u}_{\mathrm{f}}| & \theta_{\mathrm{f}}<0.8\\ \dfrac{3}{4}C_{\mathrm{D}}\dfrac{\theta_{\mathrm{s}}\rho_{\mathrm{f}}}{d_{\mathrm{s}}}|\boldsymbol{u}_{\mathrm{s}}-\boldsymbol{u}_{\mathrm{f}}|\theta_{\mathrm{f}}^{-2.65} & \theta_{\mathrm{f}}\geq 0.8\end{cases} \tag{94}$$

$$C_{\mathrm{D}}=\begin{cases}\dfrac{24\left(1+0.15Re_{\mathrm{p}}^{0.687}\right)}{Re_{\mathrm{p}}} & Re_{\mathrm{p}}\leq 1000\\ 0.44 & Re_{\mathrm{p}}>1000\end{cases} \tag{95}$$

θ_{s} is the particle volume fraction, θ_{f} is the fluid volume fraction, μ_{f} is the fluid phase viscosity, ρ_{f} is the fluid density, d_{s} is the particle diameter of the solid phase and C_{D} is the drag coefficient. Particle Reynolds number Re_{p} is described by

$$Re_{\mathrm{p}}=\theta_{\mathrm{f}}\rho_{\mathrm{f}}d_{\mathrm{s}}|\boldsymbol{u}_{\mathrm{s}}-\boldsymbol{u}_{\mathrm{f}}|/\boldsymbol{u}_{\mathrm{f}} \tag{96}$$

Therefore, Eq. (40), the momentum equation in the x direction, is adjusted by a source term as follows

$$\frac{\partial u}{\partial t}+u\frac{\partial u}{\partial x}+v\frac{\partial u}{\partial y}+w\frac{\partial u}{\partial z}=-\frac{1}{\rho}\frac{\partial p}{\partial x}+\frac{\mu}{\rho}\left(\frac{\partial^2 u}{\partial x^2}+\frac{\partial^2 u}{\partial y^2}+\frac{\partial^2 u}{\partial z^2}\right)+g_x+S_{\mathrm{mom},x} \tag{97}$$

where S_{mom} is the source term for momentum which can be expressed as:

$$S_{\mathrm{mom},x}=\sum F_{\mathrm{D},i,x}/(\rho\mathrm{d}x\mathrm{d}y\mathrm{d}z) \tag{98}$$

Similarly, the momentum equation in the y direction (Eq. 41) and the momentum equation in the z direction (Eq. 42) are adjusted as follows

$$\frac{\partial v}{\partial t}+u\frac{\partial v}{\partial x}+v\frac{\partial v}{\partial y}+w\frac{\partial v}{\partial z}=-\frac{1}{\rho}\frac{\partial p}{\partial y}+\frac{\mu}{\rho}\left(\frac{\partial^2 v}{\partial x^2}+\frac{\partial^2 v}{\partial y^2}+\frac{\partial^2 v}{\partial z^2}\right)+g_y+S_{\text{mom},y} \tag{99}$$

$$\frac{\partial w}{\partial t}+u\frac{\partial w}{\partial x}+v\frac{\partial w}{\partial y}+w\frac{\partial w}{\partial z}=-\frac{1}{\rho}\frac{\partial p}{\partial z}+\frac{\mu}{\rho}\left(\frac{\partial^2 w}{\partial x^2}+\frac{\partial^2 w}{\partial y^2}+\frac{\partial^2 w}{\partial z^2}\right)+g_z+S_{\text{mom},z} \tag{100}$$

3.4 Source term for the energy equation

As discussed in the energy equation section, the energy equation ensures the thermal energy conservation. In some cases, chemical reactions occur in the fluid domain, and these reactions are either exothermic or endothermic. If the reaction is exothermic, it could release heat to the surrounding fluid and increase the temperature. If the reaction is endothermic, it consumes the heat from the surrounding fluid and decreases the temperature. Even though total energy is conserved, the thermal energy is not. In order to consider the imbalance of thermal energy in fluid, the source term of heat must be included in the energy equation. By considering reaction heat, the energy balance is

$$\rho\frac{DE}{Dt}\Delta x\Delta y\Delta z=\sum\dot{Q}+r\Delta h\Delta x\Delta y\Delta z \tag{101}$$

where Δh (J kg^{-1}) is the heat generated when unit mass reactant is converted to products. Bringing Fourier's law of heat conduction and dividing both sides by $\Delta x\Delta y\Delta z$, the energy equation with source term $S_{\text{energ}} = r\Delta h$ is

$$\rho\frac{DE}{Dt}=\lambda\frac{\partial}{\partial x}\left(\frac{\partial T}{\partial x}\right)+\lambda\frac{\partial}{\partial y}\left(\frac{\partial T}{\partial y}\right)+\lambda\frac{\partial}{\partial z}\left(\frac{\partial T}{\partial z}\right)+r\Delta h \tag{102}$$

With the inclusion of source terms, processes with chemical reactions or multiphase interactions could be accurately simulated in CFD models.

4. Numerical methods to solve governing equations

The governing equations discussed above are almost impossible to solve analytically. The only feasible pathway to solve these partial differential equations is employing numerical methods. This is the reason most CFD software need to discretize the fluid domain to generate a mesh or grid. The core idea of numerical methods is transforming differential to difference quotient. The commonly used numerical methods in most CFD software include the finite difference method and the finite volume method.

4.1 Finite difference method

The finite difference method is the first method for solving partial differential equations. In finite difference method, the Taylor series is employed to generate approximations to the partial derivatives of the CFD governing equations. According to the Taylor series, the value of a general variable ϕ at node i + 1 (**Figure 12**) could be expressed as:

$$\phi_{i+1} = \phi_i + \left(\frac{\partial \phi}{\partial x}\right)_i \Delta x + \left(\frac{\partial^2 \phi}{\partial x^2}\right)_i \frac{\Delta x^2}{2} + \left(\frac{\partial^3 \phi}{\partial x^3}\right)_i \frac{\Delta x^3}{6} + (\text{Higher Order Terms}) \tag{103}$$

And the value of a general variable φ at node *i*-1 is:

$$\phi_{i-1} = \phi_i - \left(\frac{\partial \phi}{\partial x}\right)_i \Delta x + \left(\frac{\partial^2 \phi}{\partial x^2}\right)_i \frac{\Delta x^2}{2} - \left(\frac{\partial^3 \phi}{\partial x^3}\right)_i \frac{\Delta x^3}{6} + (\text{Higher Order Terms}) \tag{104}$$

If we subtract Eq. (104) from Eq. (103), we get

$$\left(\frac{\partial \phi}{\partial x}\right)_i = \frac{\phi_{i+1} - \phi_{i-1}}{2\Delta x} + \frac{(\Delta x^3/3)(\partial^3 \phi/\partial x^3)}{2\Delta x} \tag{105}$$

Since Δx is approaching zero, the second term is very small compared to the first term. By omitting the Δx^3 term, the central difference is achieved

$$\left(\frac{\partial \phi}{\partial x}\right)_i = \frac{\phi_{i+1} - \phi_{i-1}}{2\Delta x} \tag{106}$$

From Eq. (103) and assuming Δx is approaching zero, we can get the forward difference

$$\left(\frac{\partial \phi}{\partial x}\right)_i = \frac{\phi_{i+1} - \phi_i}{\Delta x} + O(\Delta x) \tag{107}$$

From Eq. (104) and assuming Δx is approaching to zero, we can get the backward difference

$$\left(\frac{\partial \phi}{\partial x}\right)_i = \frac{\phi_i - \phi_{i-1}}{\Delta x} + O(\Delta x) \tag{108}$$

Eqs. (106)-(108) successfully approximated the first-order derivative by difference quotient. First-order derivative is actually a slope of a variable. **Figure 13** geometrically represents the slopes by backward difference, central difference and forward difference. It is clear that the slope by central difference (blue line) is closer to the real slope (black line), so the central difference could represent the exact partial differential with a higher accuracy.

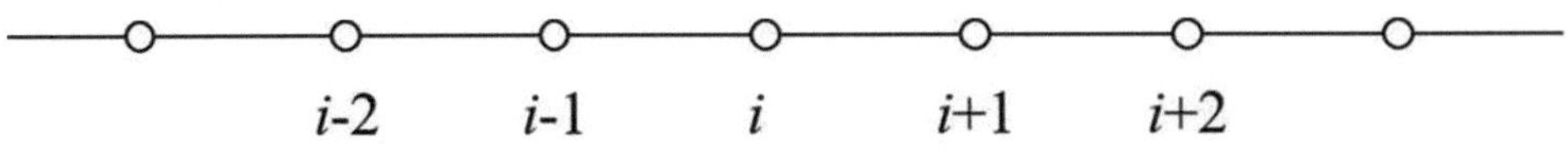

Figure 12.
One-dimensional grids for the finite-difference method.

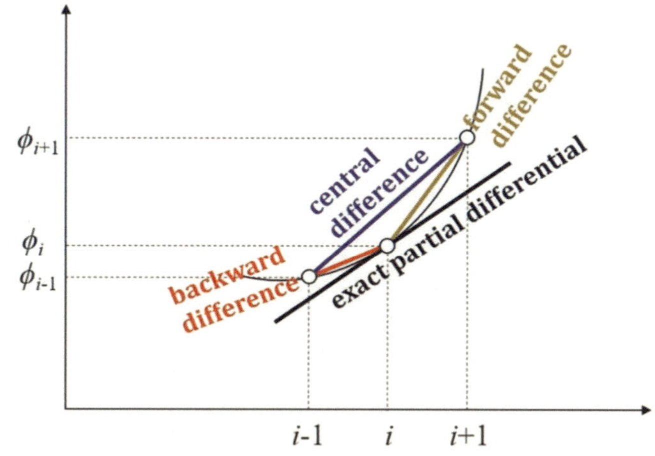

Figure 13.
Finite-difference representation of the derivative for ∂ϕ/∂x.

In CFD governing equations, there are second-order partial derivatives in diffusion terms, so the second-order derivatives should also be approximated by finite difference method. Adding Eqs. (103) and (104) together yields

$$\phi_{i+1} + \phi_{i-1} = 2\phi_i + \left(\frac{\partial^2 \phi}{\partial x^2}\right)_i \Delta x^2 + \left(\frac{\partial^4 \phi}{\partial x^4}\right)_i \frac{\Delta x^4}{24} \tag{109}$$

By omitting higher-order terms, the difference quotient form in finite difference method for second-order derivative is

$$\left(\frac{\partial^2 \phi}{\partial x^2}\right) = \frac{\phi_{i+1} - 2\phi_i + \phi_{i-1}}{\Delta x^2} \tag{110}$$

In a 3-D fluid domain such as **Figure 14**, Taylor series expansion also applies for y and z directions. The central differences for first- and second-order derivatives are:

$$\left(\frac{\partial \phi}{\partial y}\right)_j = \frac{\phi_{j+1} - \phi_{j-1}}{2\Delta y} \tag{111}$$

$$\left(\frac{\partial \phi}{\partial z}\right)_k = \frac{\phi_{k+1} - \phi_{k-1}}{2\Delta z} \tag{112}$$

$$\left(\frac{\partial^2 \phi}{\partial y^2}\right) = \frac{\phi_{j+1} - 2\phi_j + \phi_{j-1}}{\Delta y^2} \tag{113}$$

$$\left(\frac{\partial^2 \phi}{\partial z^2}\right) = \frac{\phi_{k+1} - 2\phi_k + \phi_{k-1}}{\Delta z^2} \tag{114}$$

For time term derivatives, only the first-order derivative needs to be approximated by the finite difference method. In most cases, the backward difference is used (Eq. (115)). This is because the variable value at the previous time step has already been calculated, and then, it is available to calculate the current time step in explicit expression. Since the computation performs in chronological order, the variable

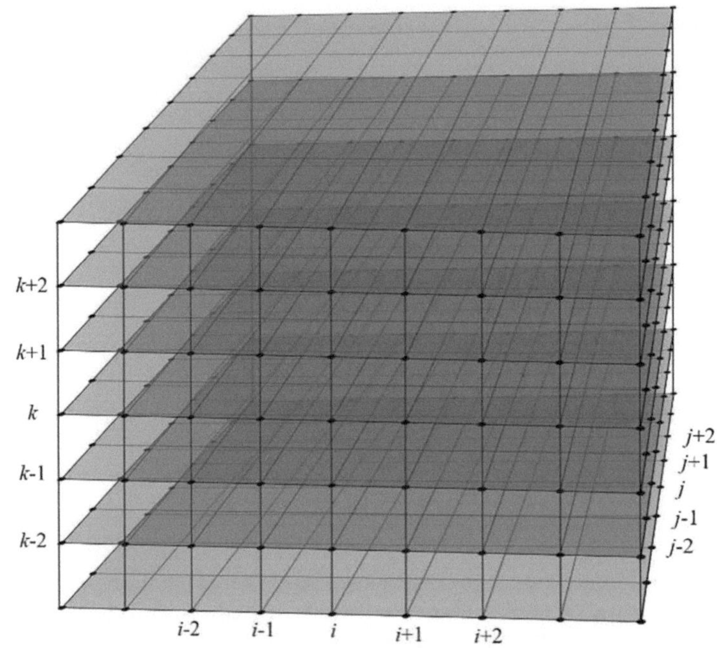

Figure 14.
Three-dimensional grids for the finite-difference method.

values at future steps are not available. Forward difference and central difference involve the variable at future steps (Eqs. (116) and (117)), so it is not recommended to be used in explicit solver. It can only be solved in an implicit solver.

$$\left(\frac{\partial\phi}{\partial t}\right) = \frac{\phi_{i,j,k}{}^{n} - \phi_{i,j,k}{}^{n-1}}{\Delta t} \tag{115}$$

$$\left(\frac{\partial\phi}{\partial t}\right) = \frac{\phi_{i,j,k}{}^{n+1} - \phi_{i,j,k}{}^{n}}{\Delta t} \tag{116}$$

$$\left(\frac{\partial\phi}{\partial t}\right) = \frac{\phi_{i,j,k}{}^{n+1} - \phi_{i,j,k}{}^{n-1}}{2\Delta t} \tag{117}$$

As we know it is easy to transform difference quotient to linear equation. With the finite element method, all the CFD partial differential equations can be converted to linear equations. By applying the techniques of solving linear equations, the numerical solutions of fluid field could be obtained.

4.2 Finite volume method

For structured mesh (**Figure 14**), it is ideal to use the finite difference method introduced above. However, most CFD simulations have to use an unstructured mesh (**Figure 15**) due to the complexity of geometries, so it is difficult to define the node sequence i, j and k. Therefore, it is almost impossible to use the finite difference method. In the case of unstructured mesh, we can resort to another popular numerical method which is called the finite volume method.

Figure 15.
Examples of unstructured mesh.

Finite volume method is derived from Gauss's divergence theorem. In divergence theorem, the integration of a variable divergence over a body equals the integration of the variable over the surface which encloses the body.

$$\begin{aligned}\iiint_{\Omega} \frac{\partial \phi_x}{\partial x} + \frac{\partial \phi_y}{\partial y} + \frac{\partial \phi_z}{\partial z} \mathrm{d}V &= \oiint_{\Sigma} \vec{\phi}(x,y,z)\mathrm{d}S \\ &= \oiint_{\Sigma} \phi_x \mathrm{d}y\mathrm{d}z + \oiint_{\Sigma} \phi_y \mathrm{d}x\mathrm{d}z + \oiint_{\Sigma} \phi_z \mathrm{d}x\mathrm{d}y\end{aligned} \tag{118}$$

It can also be written in x, y, z directions as follows

$$\iiint_{\Omega} \frac{\partial \phi}{\partial z} \mathrm{d}V = \iint_{\Sigma} \phi(x,y,z)\mathrm{d}x\mathrm{d}y \tag{119}$$

$$\iiint_{\Omega} \frac{\partial \phi}{\partial x} \mathrm{d}V = \iint_{\Sigma} \phi(x,y,z)\mathrm{d}y\mathrm{d}z \tag{120}$$

$$\iiint_{\Omega} \frac{\partial \phi}{\partial y} \mathrm{d}V = \iint_{\Sigma} \phi(x,y,z)\mathrm{d}x\mathrm{d}z \tag{121}$$

In CFD, we assume $\partial\phi/\partial x$ only has variation between different grids. Inside a single grid, it has only a constant value of $\partial\phi/\partial x$. The variation inside a single grid is very small and could be neglected. Therefore, for a grid, the left side of Gauss's divergence theorem could be simplified as

$$\iiint_{\Omega} \frac{\partial \phi}{\partial x} \mathrm{d}V = \frac{\partial \phi}{\partial x} \iiint_{\Omega} \mathrm{d}V = \frac{\partial \phi}{\partial x} \Delta V \tag{122}$$

Similarly, the variation of ϕ on a grid surface could also be neglected if we assume a grid surface is too small to have variation of ϕ. The right side of Gauss's divergence theorem could be simplified as well

$$\iint_{\Sigma} \phi \mathrm{d}y\mathrm{d}z = \phi A^x \tag{123}$$

The divergence theorem correlated the $\partial\phi/\partial x$ to ϕ, and we can utilize this correlation to get the approximation of the first-order derivative in a tetrahedral grid (**Figure 16**) as follows

$$\begin{aligned}\frac{\partial \phi}{\partial x} &= \frac{1}{\Delta V}\int_{\Delta V}\frac{\partial \phi}{\partial x}\mathrm{d}V = \frac{1}{\Delta V}\int_{A}\phi \mathrm{d}A^x \\ &= \frac{1}{\Delta V}\left(\phi_1 A_1^x + \phi_2 A_2^x + \phi_3 A_3^x + \phi_4 A_4^x\right) \\ &= \frac{1}{\Delta V}\sum_{i=1}^{N}\phi_i A_i^x\end{aligned} \tag{124}$$

where A_i^x is the projected area to the y-o-z surface.

The first-order derivative with respect to y and z can be similarly obtained as follows

$$\begin{aligned}\frac{\partial \phi}{\partial y} &= \frac{1}{\Delta V}\int_{\Delta V}\frac{\partial \phi}{\partial y}\mathrm{d}V = \frac{1}{\Delta V}\int_{A}\phi \mathrm{d}A^y \\ &= \frac{1}{\Delta V}\left(\phi_1 A_1^y + \phi_2 A_2^y + \phi_3 A_3^y + \phi_4 A_4^y\right) \\ &= \frac{1}{\Delta V}\sum_{i=1}^{N}\phi_i A_i^y\end{aligned} \tag{125}$$

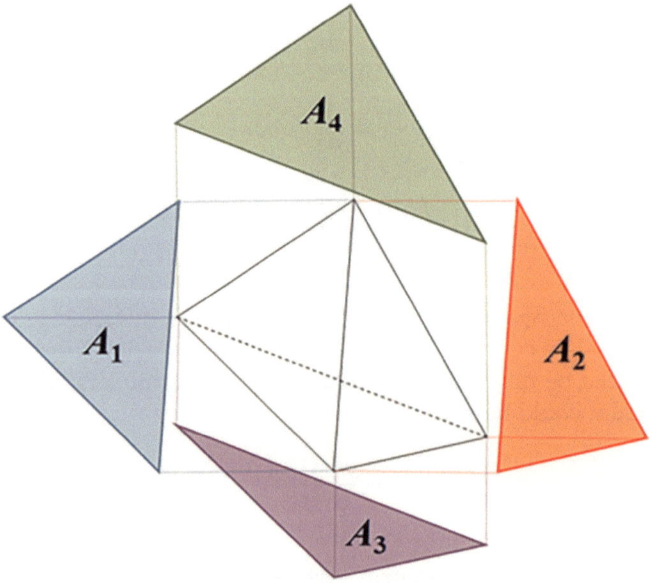

Figure 16.
The representation of a tetrahedral grid.

$$\begin{aligned}\frac{\partial\phi}{\partial z} &= \frac{1}{\Delta V}\int_{\Delta V}\frac{\partial\phi}{\partial z}\mathrm{d}V = \frac{1}{\Delta V}\int_{A}\phi\mathrm{d}A^z \\ &= \frac{1}{\Delta V}\left(\phi_1A_1^z + \phi_2A_2^z + \phi_3A_3^z + \phi_4A_4^z\right) \\ &= \frac{1}{\Delta V}\sum_{i=1}^{N}\phi_iA_i^z\end{aligned} \tag{126}$$

In the same way, the second-order derivative can be approximated by the first-order derivative

$$\begin{aligned}\frac{\partial^2\phi}{\partial x^2} &= \frac{1}{\Delta V}\int_{\Delta V}\frac{\partial^2\phi}{\partial x^2}\mathrm{d}V = \frac{1}{\Delta V}\int_{A}\frac{\partial\phi}{\partial x}\mathrm{d}A^x \\ &= \frac{1}{\Delta V}\left[\left(\frac{\partial\phi}{\partial x}\right)_1A_1^x + \left(\frac{\partial\phi}{\partial x}\right)_2A_2^x + \left(\frac{\partial\phi}{\partial x}\right)_3A_3^x + \left(\frac{\partial\phi}{\partial x}\right)_4A_4^x\right] \\ &= \frac{1}{\Delta V}\sum_{i=1}^{N}\left(\frac{\partial\phi}{\partial x}\right)_iA_i^x\end{aligned} \tag{127}$$

$$\frac{\partial^2\phi}{\partial y^2} = \frac{1}{\Delta V}\sum_{i=1}^{N}\left(\frac{\partial\phi}{\partial y}\right)_iA_i^y \tag{128}$$

$$\frac{\partial^2\phi}{\partial z^2} = \frac{1}{\Delta V}\sum_{i=1}^{N}\left(\frac{\partial\phi}{\partial z}\right)_iA_i^z \tag{129}$$

From the equations above, it is clear that the finite volume method has better conservativeness of the CFD governing equations. Meanwhile it is more adaptive to complicated geometries. In most CFD commercial software, the finite volume method is the mainstream technique to solve the CFD governing equations.

5. Conclusion

Understanding the basic theory of CFD governing equations and numerical solutions is the key to developing CFD simulation codes as well as using CFD packages. Basically, all flows can be universally described by the continuity equation, component equation, momentum equation and energy equation, because fluid follows mass conservation, momentum conservation and energy conservation. With source terms, CFD applications were broadened to a wider field like chemical engineering, environmental engineering or thermal engineering. However, these equations are almost impossible to be solved analytically, thus we have to resort to the numerical method to get the approximate results. This involves using the finite difference method and finite volume method. The numerical methods offered a feasible way to get the solutions from the governing equations, which facilitated the analysis of complex phenomena associated with fluid flow, heat transfer and chemical reactions.

Acknowledgements

Authors appreciate the financial support from teaching reform fund for postgraduate students in Dalian University of Technology (No. 20), Liao Ning Revitalization Talents Program (grant number: XLYC2007179) and the Australian Research Council (grant number DE210101549).

Author details

Guozhao Ji[1], Meng Zhang[1,2], Yongming Lu[1] and Jingliang Dong[3]*

1 Key Laboratory of Industrial Ecology and Environmental Engineering (MOE), School of Environmental Science and Technology, Dalian University of Technology, Dalian, Liaoning, China

2 Mechanical and Automotive Engineering, School of Engineering, RMIT University, Bundoora, Victoria, Australia

3 Institute for Sustainable Industries and Liveable Cities, Victoria University, Melbourne, Victoria, Australia

*Address all correspondence to: jingliang.dong@vu.edu.au

References

[1] Hasse C, Debiagi P, Wen X, Hildebrandt K, Vascellari M, Faravelli T. Advanced modeling approaches for CFD simulations of coal combustion and gasification. Progress in Energy and Combustion Science. 2021;**86**:100938. DOI: 10.1016/j.pecs.2021.100938

[2] Ramos A, Monteiro E, Rouboa A. Numerical approaches and comprehensive models for gasification process: A review. Renewable and Sustainable Energy Reviews. 2019;**110**: 188-206. DOI: 10.1016/j.rser.2019.04.048

[3] Tu Q, Wang H, Ocone R. Application of three-dimensional full-loop CFD simulation in circulating fluidized bed combustion reactors – A review. Powder Technology. 2022;**399**:117181. DOI: 10.1016/j.powtec.2022.117181

[4] Luo H, Wang X, Liu X, Wu X, Shi X, Xiong Q. A review on CFD simulation of biomass pyrolysis in fluidized bed reactors with emphasis on particle-scale models. Journal of Analytical and Applied Pyrolysis. 2022;**162**:105433. DOI: 10.1016/j.jaap.2022.105433

[5] Lu L, Gao X, Dietiker J-F, Shahnam M, Rogers WA. MFiX based multi-scale CFD simulations of biomass fast pyrolysis: A review. Chemical Engineering Science. 2022;**248**:117131. DOI: 10.1016/j.ces.2021.117131

[6] Xiong Q, Yang Y, Xu F, Pan Y, Zhang J, Hong K, et al. Overview of computational fluid dynamics simulation of reactor-scale biomass pyrolysis. ACS Sustainable Chemistry & Engineering. 2017;**5**(4):2783-2798. DOI: 10.1021/acssuschemeng.6b02634

[7] Ku X, Li T, Løvås T. CFD–DEM simulation of biomass gasification with steam in a fluidized bed reactor. Chemical Engineering Science. 2015;**122**:270-283. DOI: 10.1016/j.ces.2014.08.045

[8] Ji G, Zhao M, Wang G. Computational fluid dynamic simulation of a sorption-enhanced palladium membrane reactor for enhancing hydrogen production from methane steam reforming. Energy. 2018;**147**: 884-895. DOI: 10.1016/j.energy.2018.01.092

[9] Ji G, Wang G, Hooman K, Bhatia S, Diniz da Costa JC. Computational fluid dynamics applied to high temperature hydrogen separation membranes. Frontiers of Chemical Science and Engineering. 2012;**6**(1):3-12. DOI: 10.1007/s11705-011-1161-5

[10] Ji G, Wang G, Hooman K, Bhatia S, Diniz da Costa JC. The fluid dynamic effect on the driving force for a cobalt oxide silica membrane module at high temperatures. Chemical Engineering Science. 2014;**111**:142-152. DOI: 10.1016/j.ces.2014.02.006

[11] Khan M, Hussain A, Malik MY, Salahuddin T, Aly S. Numerical analysis of Carreau fluid flow for generalized Fourier's and Fick's laws. Applied Numerical Mathematics. 2019;**144**: 100-117. DOI: 10.1016/j.apnum.2019.05.018

[12] Ji G, Wang G, Hooman K, Bhatia S, Diniz da Costa JC. Simulation of binary gas separation through multi-tube molecular sieving membranes at high temperatures. Chemical Engineering Journal. 2013;**218**:394-404. DOI: 10.1016/j.cej.2012.12.063

[13] Kundu PK, Cohen IM, Dowling DR. Fluid Mechanics. Academic Press; 2015 Amsterdam

[14] Bernardin C, Olla S. Fourier's law for a microscopic model of heat conduction. Journal of Statistical Physics. 2005; **121**(3):271-289. DOI: 10.1007/s10955-005-7578-9

[15] Zhang M, Zhang Y, Ma D, Li A, Fu W, Ji G, et al. Numerical investigation on the heat transfer of plastic waste pyrolysis in a rotary furnace. Chemical Engineering Journal. 2022;**445**:136686. DOI: 10.1016/j.cej.2022.136686

Chapter 2

Hyperbolic Navier-Stokes with Reconstructed Discontinuous Galerkin Method

Lingquan Li and Jialin Lou

Abstract

One of the crucial issues of computational fluid dynamics is how to discretize the viscous terms accurately. Recently, an attractive and viable alternative numerical method for solving the compressible Navier–Stokes equations is proposed. The first-order hyperbolic system (FOHS) with reconstructed discontinuous Galerkin (rDG) method was first proposed to solve advection–diffusion model equations and then extend to compressible Navier–Stokes equations. For the model advection–diffusion equation, the proposed method is reliable, accurate, efficient, and robust, benefiting from FOHS and rDG methods. To implement the method of compressible Navier–Stokes equations, the gradients of density, velocity, and temperature are introduced as auxiliary variables. Numerical experiments demonstrate that the developed HNS + rDG methods are able to achieve the designed order of accuracy for both primary variables and their gradients.

Keywords: reconstructed discontinuous Galerkin, hyperbolic Navier–Stokes, first-order hyperbolic system, integration methods, laminar flow

1. Introduction

The compressible Navier–Stokes equations are useful because they describe the physics of many phenomena of scientific and engineering interests. They are widely used but not limited to the fields of fluid dynamics, aeronautical engineering, and aerodynamics engineering. The Navier–Stokes equations are certain partial differential equations that describe the motion of viscous fluid substances. The compressible Navier–Stokes equations mathematically express the continuity, the theorem of momentum, and the theorem of energy of Newtonian fluids. These equations arise from applying Isaac Newton's second law to fluid motion, with the assumption that the stress in the fluid is the sum of a diffusing viscous term and a pressure term. The diffusing viscous term is proportional to the gradient of velocity. The difference between Euler equations and Navier–Stokes equations is that Euler equations are only valid for inviscid flow, while Navier–Stokes equations take viscosity into account. Euler equations are hyperbolic, while Navier–Stokes equations are parabolic.

Although the discontinuous Galerkin (DG) method is a natural choice for hyperbolic problems, it becomes far less certain when it comes to elliptic or parabolic problems. An alternative approach for viscous discretization, which reformulates the viscous terms as a first-order hyperbolic system (FOHS), was given in Refs. [1, 2]. Another FOHS formulation [3, 4] was developed by including the gradient quantities as additional variables. Over several years of development, the FOHS method has been shown to offer several distinguished advantages. First, the well-developed numerical schemes for hyperbolic systems can be directly applied to viscous problems. Second, it enables high-quality gradient prediction even on fully irregular grids. This feature shows its significance when it comes to viscous flow simulations on complex grids, where the qualities of the meshes are likely to be highly irregular. Third, due to the fact that the second-order derivatives in the governing partial differential equations (PDE) are replaced and eliminated, the developed scheme shows speedup and robustness for the iterative solvers. Finally, the FOHS method can improve viscous discretization as well as inviscid discretization. Due to these favorable characteristics, the hyperbolic methods have been implemented in various applications, including diffusion [5], anisotropic diffusion [6], advection–diffusion [7], Navier-Stokes (NS) equations [8], and three-dimensional compressible NS equations with proper handling of high Reynolds number boundary layer flows [8].

To make it easier for the readers to understand, the classical computation methods for fluid dynamics are first discussed, and the reconstructed discontinuous Galerkin (rDG) method is introduced. FOHS for linear advection–diffusion equation is then given. Eventually, the hyperbolic Navier–Stokes with rDG method is given.

2. Integration methods for first- and second-order operators

Computational fluid dynamics (CFD) is, in part, the art of replacing the governing partial differential equations of fluid flow with numbers and advancing these numbers in space and time to obtain a final numerical description of the flow [9]. To achieve this goal, accurate spatial and temporal discretization becomes necessary. CFD methods on unstructured grids mainly fall into three categories: finite element (FE) methods, finite volume (FV) methods, and DG methods.

2.1 Finite volume methods

Finite volume methods assume that the underlying solution is constant in each cell, and a Riemann problem [10] arises as discontinuity occurs at the boundary where two elements are adjacent to each other. For first-order operator

$$\nabla \cdot \mathbf{f}(u), \tag{1}$$

the finite volume method gives the formulation as

$$\int_{\Omega_{el}} \nabla \cdot \mathbf{f}(u) d\Omega = -\int_{\Gamma_{el}} \mathbf{n} \cdot \mathbf{f}(u) d\Gamma, \tag{2}$$

where Ω_{el} is the volume of the element, and Γ_{el} is the boundary faces of the element. This implies that only the normal fluxes through the element faces $\mathbf{n} \cdot \mathbf{f}(u)$ appear in the discretization. For operators with second-order derivatives, the integration is no longer obvious [11]. A number of strategies have been devised to circumvent this limitation. One of the popular methods is to evaluate the first-order derivatives in the first pass over the mesh and then obtain the second-order derivatives in a subsequent pass. A new hyperbolic Navier–Stokes (HNS) formulation is given in Ref. [12], where the gradients of density, velocity, and temperature are introduced as auxiliary variables. Therefore, the second-order derivatives can be integrated like any conservative variables in NS equations.

2.2 Finite element methods

Finite element methods are classically used for elliptic or parabolic problems. In the finite element method, a given domain is viewed as a collection of sub-domains, and over each sub-domain. Then, it becomes easier to represent a complicated function as a collection of simple polynomials [13]. As shown in **Figure 1**, the underlying solution is continuous at the element boundary. For second-order operator

$$\nabla \cdot (\nabla u), \tag{3}$$

FE formulation gives

$$\int_{\Omega} w\nabla \cdot (\nabla u) d\Omega = \int_{\Omega} (\nabla u) \cdot (\nabla w) d\Omega - \int_{\Gamma} w\mathbf{n} \cdot (\nabla u) d\Gamma, \tag{4}$$

where w is a set of test functions. Volume integration is performed in the entire computation domain Ω, and boundary integration is performed at the boundary Γ of the entire domain. In **Figure 1**, Ω includes elements 1–6. Γ includes the left boundary of element 1 and the right boundary of element 6.

2.3 Discontinuous Galerkin methods

DG methods combine the advantages of finite volume and finite element methods. While assuming the underlying solution to be polynomial on each element, the Riemann problem is solved at the element boundary. For the first-order operator (Eq. (1)), the DG formulation gives

$$\int_{\Omega_{el}} w\nabla \cdot \mathbf{f}(u) d\Omega = \int_{\Omega_{el}} \mathbf{f}(u) \cdot (\nabla w) d\Omega - \int_{\Gamma_{el}} w\mathbf{n} \cdot \mathbf{f}(u) d\Gamma, \tag{5}$$

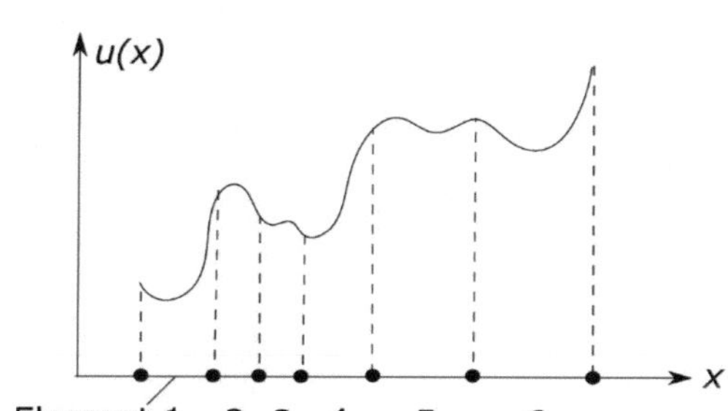

Figure 1.
Piece-wise approximation of a function as $u(x) = \sum_{el=1}^{6} u_{el}(x)$ in finite element methods.

In practice, w can be easily taken as the polynomial basis of the underlying DG solution. For second-order operator (Eq. (3)), DG formulation gives

$$\int_{\Omega_{el}} w\nabla\cdot(\nabla u)d\Omega = \int_{\Omega_{el}} (\nabla u)\cdot(\nabla w)d\Omega - \int_{\Gamma_{el}} w\mathbf{n}\cdot(\nabla u)d\Gamma. \tag{6}$$

As the DG method assumes that the solution is distributed as a polynomial function on each element, the value of ∇u is given explicitly on Ω_{el}. However, ∇u becomes discontinuous at the element boundary Γ_{el}. The flux can be estimated with BR1 method [14], BR2 method [15], or direct discontinuous Galerkin (DDG) method [16, 17]. One can achieve high-order accuracy while retaining the compactness of the stencil. Meanwhile, DG methods are especially suitable for hyperbolic systems, treatment of nonconforming meshes, and implementation of the hp-adaptive method. However, the DG method suffers from a number of weaknesses. In particular, how to reduce computing costs and how to discretize and efficiently solve elliptic/parabolic equations remain challenging issues in DG methods [7].

2.4 Reconstructed discontinuous Galerkin methods

In rDG methods, a higher-order solution is reconstructed based on the underlying solution. (Eq. (5)) becomes

$$\int_{\Omega_{el}} \mathbf{f}\left(u^R\right)\cdot(\nabla w)d\Omega - \int_{\Gamma_{el}} w\mathbf{n}\cdot\mathbf{f}\left(u^R\right)d\Gamma, \tag{7}$$

and (Eq. (6)) becomes

$$\int_{\Omega_{el}} \left(\nabla u^R\right)\cdot(\nabla w)d\Omega - \int_{\Gamma_{el}} w\mathbf{n}\cdot\left(\nabla u^R\right)d\Gamma, \tag{8}$$

where u^R is a higher-order solution reconstructed from the underlying solution u.

3. FOHS for linear advection: diffusion equation

The two-dimension linear advection–diffusion equation is written as

$$\frac{\partial\phi}{\partial t} + a\frac{\partial\phi}{\partial x} + b\frac{\partial\phi}{\partial y} = \nu\left(\frac{\partial^2\phi}{\partial x^2} + \frac{\partial^2\phi}{\partial y^2}\right) + f(x,y), \tag{9}$$

where ϕ is a scalar function, which can be referred to as the primary solution variable, or the velocity potential. (a, b) is a constant advection vector, ν is a positive diffusion coefficient, and $f(x, y)$ is the source term. The first-order derivatives of ϕ are introduced as additional variables

$$v_x = \frac{\partial\phi}{\partial x}, \quad v_y = \frac{\partial\phi}{\partial y}. \tag{10}$$

By adding pseudo time (τ) derivatives with respect to both ϕ and additional variables, the following FOHS can be formulated.

$$\begin{cases} \dfrac{\partial \phi}{\partial \tau} + \dfrac{\partial \phi}{\partial t} + a\dfrac{\partial \phi}{\partial x} + b\dfrac{\partial \phi}{\partial y} = \nu\left(\dfrac{\partial v_x}{\partial x} + \dfrac{\partial v_y}{\partial y}\right) + f(x,y), \\ \dfrac{\partial v_x}{\partial \tau} = \dfrac{1}{T_r}\left(\dfrac{\partial \phi}{\partial x} - v_x\right), \\ \dfrac{\partial v_y}{\partial \tau} = \dfrac{1}{T_r}\left(\dfrac{\partial \phi}{\partial y} - v_y\right), \end{cases} \tag{11}$$

where t and τ are the physical time and the pseudo-time, respectively. If the steady solution in pseudo-time is obtained, then the extra variables would relax to the first-order derivatives of ϕ. T_r is the relaxation time, which is a free parameter. System (Eq. (11)) is equivalent (Eq. (9)) in the pseudo-steady state for any nonzero T_r, but T_r needs to be positive for the system to be hyperbolic. The FOHS can be written in vector form as

$$\frac{\partial \mathbf{U}}{\partial \tau} + \mathbf{T}\frac{\partial \mathbf{U}}{\partial t} + \frac{\partial \mathbf{F}_x}{\partial x} + \frac{\partial \mathbf{F}_y}{\partial y} = \mathbf{S}, \tag{12}$$

where

$$\mathbf{U} = \begin{pmatrix} \phi \\ v_x \\ v_y \end{pmatrix}, \quad \mathbf{T} = \begin{pmatrix} 1 & 0 & 0 \\ 0 & 0 & 0 \\ 0 & 0 & 0 \end{pmatrix}, \tag{13}$$

$$\mathbf{F}_x = \begin{pmatrix} a\phi - \nu v_x \\ -\phi/T_r \\ 0 \end{pmatrix}, \quad \mathbf{F}_y = \begin{pmatrix} b\phi - \nu v_y \\ 0 \\ -\phi/T_r \end{pmatrix}, \quad \mathbf{S} = \begin{pmatrix} f(x,y) \\ -v_x/T_r \\ -v_y/T_r \end{pmatrix}. \tag{14}$$

To simplify the mathematics, the advection term and the diffusive term are considered separately.

$$\mathbf{F}_x = \mathbf{F}_x^a + \mathbf{F}_x^d = \begin{pmatrix} a\phi \\ 0 \\ 0 \end{pmatrix} + \begin{pmatrix} -\nu v_x \\ -\phi/T_r \\ 0 \end{pmatrix}, \tag{15}$$

$$\mathbf{F}_y = \mathbf{F}_y^a + \mathbf{F}_y^d = \begin{pmatrix} b\phi \\ 0 \\ 0 \end{pmatrix} + \begin{pmatrix} -\nu v_y \\ 0 \\ -\phi/T_r \end{pmatrix}. \tag{16}$$

Consider the Jacobian of the flux projected along $\mathbf{n} = (n_x, n_y)$,

$$\mathbf{A}_n = \mathbf{A}_n^a + \mathbf{A}_n^d = \frac{\partial \mathbf{F}_x}{\partial \mathbf{U}} n_x + \frac{\partial \mathbf{F}_y}{\partial \mathbf{U}} n_y, \tag{17}$$

where $\mathbf{A}_n^a$ and $\mathbf{A}_n^d$ are the advective and diffusive Jacobians, respectively.

$$\mathbf{A}_n^a = \frac{\partial \mathbf{F}_x^a}{\partial \mathbf{U}} n_x + \frac{\partial \mathbf{F}_y^a}{\partial \mathbf{U}} n_y = \begin{pmatrix} a_n & 0 & 0 \\ 0 & 0 & 0 \\ 0 & 0 & 0 \end{pmatrix}, \tag{18}$$

$$\mathbf{A}_n^d = \frac{\partial \mathbf{F}_x^d}{\partial \mathbf{U}} n_x + \frac{\partial \mathbf{F}_y^d}{\partial \mathbf{U}} n_y = \begin{pmatrix} 0 & -\nu n_x & -\nu n_y \\ -n_x/T_r & 0 & 0 \\ -n_y/T_r & 0 & 0 \end{pmatrix}, \tag{19}$$

where

$$a_n = a n_x + b n_y. \tag{20}$$

The only nonzero eigenvalue of the advective Jacobian is a_n, while the eigenvalues of the diffusive Jacobian are

$$\lambda_1 = \sqrt{\frac{\nu}{T_r}}, \quad \lambda_2 = -\sqrt{\frac{\nu}{T_r}}, \quad \lambda_3 = 0. \tag{21}$$

One can see that the diffusion progress can be regarded as a wave propagating isotropically according to the first two nonzero eigenvalues. As for the third eigenvalue, it indicates the inconsistency damping mode [18]. Clearly, the steady solution is independent of the free parameter, relaxation time T_r. Therefore, T_r can be defined solely to accelerate the convergence. For simplicity, T_r is taken as

$$T_r = \frac{L_r^2}{\nu}, \quad L_r = \frac{1}{max\,(Re\,, 2\pi)}, \quad Re = \frac{\sqrt{a^2 + b^2}}{\nu}. \tag{22}$$

One can see that only first-order operators occur in (Eq. (12)). FV formulation (Eq. (2)), DG formulation (Eq. (5)), or rDG formulation (Eq. (7)) can be used to integrate the first-order spatial operators.

4. FOHS for Navier-Stokes equations

The FOHS method is straightforward for model equations by introducing the gradients of the unknowns as auxiliary variables. However, for compressible NS equations, the construction of the FOHS becomes trickier. A HNS14 method is first introduced by including viscous stresses and heat fluxed as additional variables [19], but it turns out to obtain reduced order of accuracy in velocity gradients. Later, a HNS 17 method is developed to improve the scheme by including the velocity gradients scaled by viscosity and heat fluxes [20], resulting in the designed order of accuracy in velocity gradients. However, the loss of designed accuracy in density gradients is observed. In order to overcome this issue, an artificial hyperbolic density diffusion term is added to HNS17 to construct HNS20 [12]. The additional term is designed to be small enough such that the scheme can provide expected accurate gradients for density while the continuity equation is not affected. At this point, accurate gradients can be obtained for all primary variables by HNS20 in the pseudo-steady state. However, this efficient construction is not straightforward for other conservative or primary variables. To make the recycling procedure trivial, a new formulation HNS20G [21] was developed recently. Unlike the above-mentioned methods, HNS20G uses the gradients of the primary variables, that is, density, velocity, and temperature, as auxiliary variables to the first-order hyperbolic system. Moreover, with the utilization of the reconstruction method, HNS20G is able to deliver a more accurate solution and gradient while remaining the same degrees of freedom as the conventional DG (P1) method.

4.1 Navier-Stokes equations

The non-dimensionalized Navier–Stokes equations governing unsteady compressible viscous flows can be expressed as

$$\begin{cases} \dfrac{\partial \rho}{\partial t} + \dfrac{\partial \rho v_k}{\partial x_k} = 0, \\ \dfrac{\rho v_i}{\partial t} + \dfrac{\partial(\rho v_i v_k + p\boldsymbol{\delta}_{ik} - \tau_{ik})}{\partial x_k} = 0, \\ \dfrac{\rho e}{\partial t} + \dfrac{\partial\left[v_j \rho e + v_i(p\delta_{ik} - \tau_{ik}) + q_k\right]}{\partial x_k} = 0, \end{cases} \tag{23}$$

where the summation convention (k = 1, 2, 3) has been used as $x_1 = x, x_2 = y$, and $x_3 = z$. The symbols ρ, p, and e denote the density, pressure, and specific total energy of the fluid, respectively, and $v_x = u, v_y = v$, and $v_z = w$ are the velocity components of the flow in the coordinate direction x, y, and z, *respectively*. The symbol $\boldsymbol{\delta}_{ik}$ is the Kronecker delta function. The pressure can be computed from the equation of state for a perfect gas,

$$p = (\gamma - 1)\rho\left(e - \frac{1}{2}\left(u^2 + v^2 + w^2\right)\right), \tag{24}$$

where the ratio of the specific heats, is assumed to be constant, that is, $\gamma = 1.4$. Moreover, the specific total enthalpy h is defined as

$$h = e + \frac{p}{\rho}. \tag{25}$$

The viscous stress tensor $\boldsymbol{\tau}$ can be found through

$$\boldsymbol{\tau} = \begin{pmatrix} \tau_{xx} & \tau_{xy} & \tau_{xz} \\ \tau_{yx} & \tau_{yy} & \tau_{yz} \\ \tau_{zx} & \tau_{zy} & \tau_{zz} \end{pmatrix}. \tag{26}$$

The Newtonian fluid with the Stokes hypothesis is valid under the current framework. Thus, $\boldsymbol{\tau}$ is a symmetric tensor, which is a linear function of the velocity gradients

$$\tau_{ij} = \mu\left(\frac{\partial u_i}{\partial x_j} + \frac{\partial u_j}{\partial x_i}\right) - \frac{2}{3}\mu\frac{\partial u_k}{\partial x_k}\boldsymbol{\delta}_{ij}, \tag{27}$$

where μ is the molecular viscosity coefficient or dynamic viscosity as in many literatures, which can be determined through Sutherland's law

$$\frac{\mu}{\mu_0} = \left(\frac{T}{T_0}\right)^{\frac{3}{2}}\frac{T_0 + S}{T + S}, \tag{28}$$

where μ_0 represents the viscosity coefficient at the reference temperature T_0 and $S = 110K$. The temperature of the fluid T is determined by

$$T = \frac{P}{\rho R}, \tag{29}$$

where R denotes the universal gas constant for a perfect gas.

According to Fourier's law, the heat flux vector q_j is given by

$$q_j = -\lambda \frac{\partial T}{\partial x_j}, \tag{30}$$

where λ is the thermal conductivity coefficient and expressed as

$$\lambda = \frac{\mu c_p}{\mathrm{Pr}}, \tag{31}$$

where c_p is the specific heat capacity at constant pressure and Pr is the nondimensional laminar Prandtl number, which is taken as 0.7 for air.

4.2 First-order hyperbolic Navier–Stokes system: HNS20G

The gradients of density, velocity, and temperature are first introduced as auxiliary variables

$$\begin{aligned} &\mathbf{r} = \begin{pmatrix} r_x \\ r_y \\ r_z \end{pmatrix} = \nabla\rho, \quad \mathbf{g}_u = \begin{pmatrix} g_{ux} \\ g_{uy} \\ g_{uz} \end{pmatrix} = \nabla u, \quad \mathbf{g}_v = \begin{pmatrix} g_{vx} \\ g_{vy} \\ g_{vz} \end{pmatrix} = \nabla v, \\ &\mathbf{g}_w = \begin{pmatrix} g_{wx} \\ g_{wy} \\ g_{wz} \end{pmatrix} = \nabla w, \quad \mathbf{h} = \begin{pmatrix} h_x \\ h_y \\ h_z \end{pmatrix} = \nabla T. \end{aligned} \tag{32}$$

Then, the governing equations of compressible viscous flows are reformulated as

$$\begin{cases} \dfrac{\partial \rho}{\partial t} + \dfrac{\partial \rho v_k - \nu_r r_k}{\partial x_k} = 0, \\ \dfrac{\rho v_i}{\partial t} + \dfrac{\partial(\rho v_i v_k + p\delta_{ik} - \tau_{ik})}{\partial x_k} = 0, \\ \dfrac{\rho e}{\partial t} + \dfrac{\partial\left[v_j \rho e + v_i(p\delta_{ik} - \tau_{ik}) + q_k\right]}{\partial x_k} = 0, \\ -\dfrac{\partial v_i}{\partial x_k} = -g_{v_i k}, \\ -\dfrac{\partial T}{\partial x_k} = -h_k, \\ -\dfrac{\partial \rho}{\partial x_k} = -r_k, \end{cases} \tag{33}$$

where ν_r is introduced as an additional term to the continuity equation as artificial viscosity associated with the artificial hyperbolic mass diffusion. In the presented work, we take $\nu_r = \min\left(10^{-12}, h^3\right)$, where h is the scale of cell size. A pseudo time τ is then introduced into Eq. (33) to make the system hyperbolic, and the resulting hyperbolic system is written as

$$\begin{cases} \dfrac{\partial \rho}{\partial \tau} + \dfrac{\partial \rho}{\partial t} + \dfrac{\partial \rho v_k - \nu_r r_k}{\partial x_k} = 0, \\ \dfrac{\partial \rho v_i}{\partial \tau} + \dfrac{\partial \rho v_i}{\partial t} + \dfrac{\partial (\rho v_i v_k + p\delta_{ik} - \tau_{ik})}{\partial x_k} = 0, \\ \dfrac{\partial \rho e}{\partial \tau} + \dfrac{\partial \rho e}{\partial t} + \dfrac{\partial \left[v_j \rho e + v_i (p\delta_{ik} - \tau_{ik}) + q_k\right]}{\partial x_k} = 0, \\ T_v \dfrac{\partial g_{v_i k}}{\partial \tau} - \dfrac{\partial v_i}{\partial x_k} = -g_{v_i k}, \\ T_h \dfrac{\partial h_k}{\partial \tau} - \dfrac{\partial T}{\partial x_k} = -h_k, \\ T_r \dfrac{\partial r_k}{\partial \tau} - \dfrac{\partial \rho}{\partial x_k} = -r_k. \end{cases} \tag{34}$$

The relaxation times T_r, T_v, and T_h are defined as

$$T_r = \frac{L_r^2}{\nu_r}, \quad T_v = \frac{L_v^2}{\nu_v}, \quad T_h = \frac{L_h^2}{\nu_h}, \tag{35}$$

where the length scale is taken as $L_r = L_v = L_h = 1/2\pi$. The normalized viscosity coefficients are

$$\nu_v = \frac{\mu_v}{\rho}, \quad \nu_h = \frac{\mu_h}{\rho}, \tag{36}$$

Note that for high Reynolds number flows, the length scale should be modified according to the Reynolds number as

$$\begin{aligned} L_r &= \frac{L_d}{\sqrt{Re_{L_d}^{r\infty}}}, \quad Re_{L_d}^{r\infty} = \frac{U_\infty L_d}{\nu_r}, \\ L_v &= \frac{L_d}{\sqrt{Re_{L_d}^{v\infty}}}, \quad Re_{L_d}^{v\infty} = \frac{U_\infty L_d}{\nu_{v\infty}}, \\ L_h &= \frac{L_d}{\sqrt{Re_{L_d}^{h\infty}}}, \quad Re_{L_d}^{h\infty} = \frac{U_\infty L_d}{\nu_{h\infty}}, \\ L_d &= \frac{1}{2\pi}, \end{aligned} \tag{37}$$

where U_∞ is the free-stream flow velocity, and the superscript and subscript ∞ refers to the free-stream flow.

After writing (Eq. (34)) in the same formulation as (Eq. (12)), either FV formulation (Eq. (2)), DG formulation (Eq. (5)), or rDG formulation (Eq. (7)) can be used to integrate the first-order spatial operators.

4.3 Laminar flow past a sphere

The laminar flow past a sphere is given here to compare the present method with experimental data. The free-stream Reynolds number is taken as $Re_\infty = 100$, and the free-stream Mach number is taken as $M_\infty = 0.5$. The triangular meshes for the symmetric plane and the spherical surface are shown in **Figures 2** and **3**. Characteristic condition is given at the left, right, upper, and lower boundaries in **Figure 2**. Adiabatic

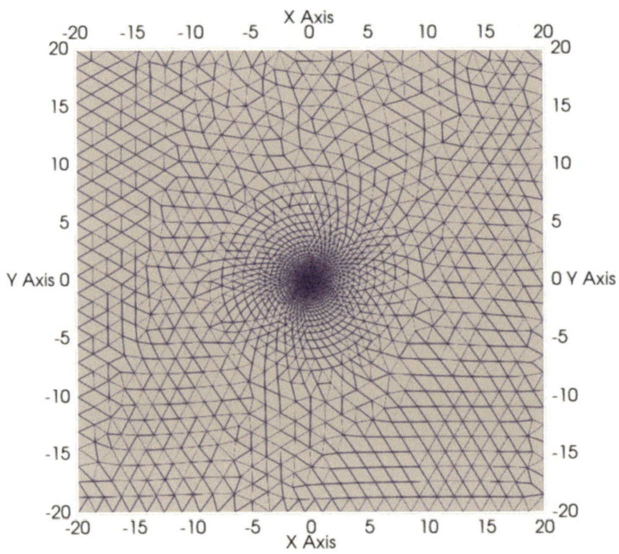

Figure 2.
Triangular meshes of the symmetric plane for laminar past a sphere.

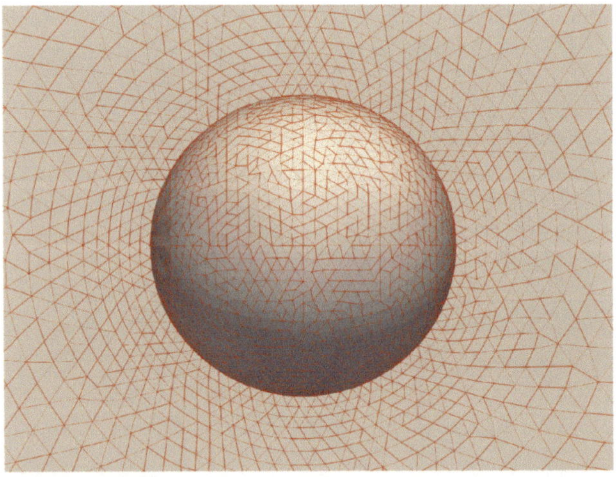

Figure 3.
Triangular meshes of the sphere surface.

wall boundary is given at the spherical surface. The vorticity magnitude contours are given in **Figure 4**. The pressure coefficient distribution on the spherical surface is compared to Ref. [22], as shown in **Figure 5**.

4.4 Laminar flow past a delta wing

A laminar flow at a high angle of attack past a delta wing is considered here. The free-stream Mach number is of $M_\infty = 0.3$, the angle of attack is of $\alpha = 12.5^o$, and the Reynolds number is of $\mathrm{Re} = 4{,}000$ based on a mean cord length of 1. The triangular

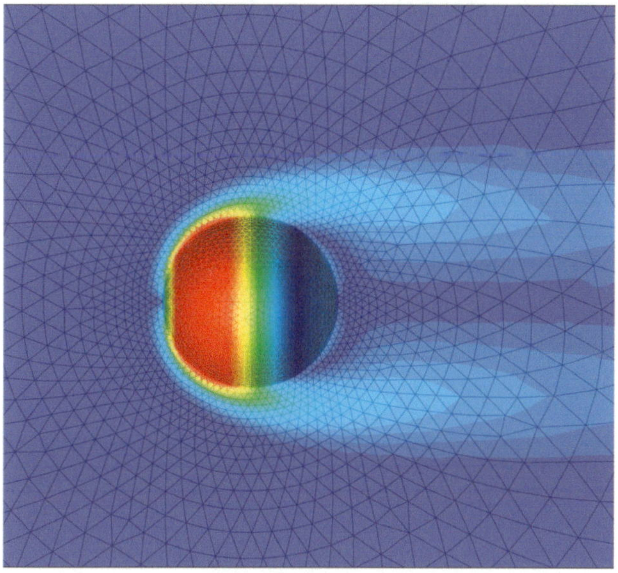

Figure 4.
Vorticity magnitude contours and for laminar past a sphere at $Re_\infty = 100$.

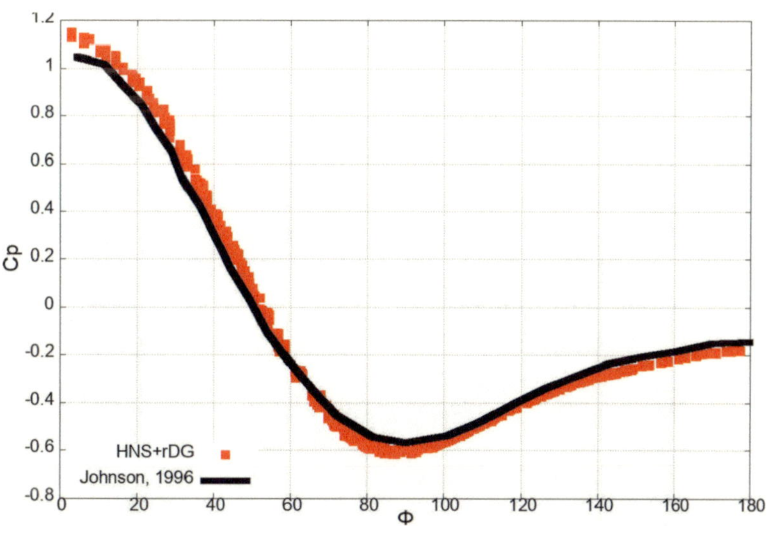

Figure 5.
Pressure coefficient for laminar past a sphere, compared to ref. [22].

meshes of the symmetric plane and the delta wing surface are shown in **Figures 6** and 7. Characteristic condition is given at the upper, lower, left, and right boundaries in **Figure 6**. The adiabatic wall boundary condition is given at the wing surface, where the heat transfer through the wall is zero, and no mass or total energy convection is supposed. The computed vorticity magnitude contours are shown in **Figure 8** along with the stream traces in the flow field. It is observed that as the flow passes the leading edge, it rolls up and creates a vortex. The vortex system remains over a distance behind the wing.

4.5 Von Kármán vortex street behind a circular cylinder

The von Kármán vortex street is one of the most extensively studied cases both experimentally and numerically in fluid dynamics. The initial condition is a uniform-free stream. The free-stream Mach number is taken as $M_\infty = 0.1$ and the Reynolds number is taken as $\mathrm{Re} = 100$ based on the diameter of the cylinder. The grid is composed of 21, 809 prismatic elements, 22, 804 grid points, 43, 168 triangular boundary faces, and 199 quadratic boundary faces, as shown in **Figures 9** and **10**. The vorticity contours within the half cycle are shown in **Figure 11**.

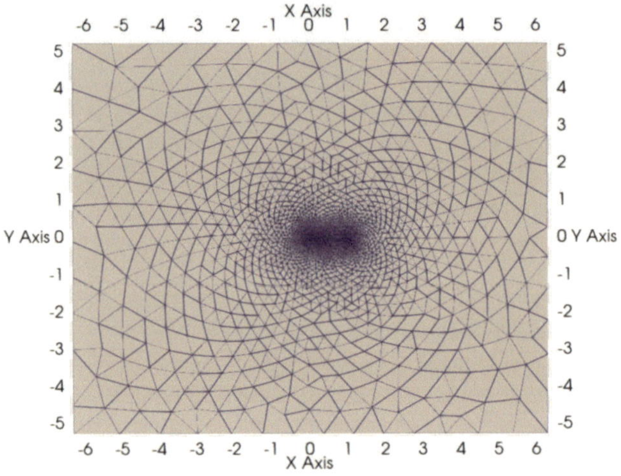

Figure 6.
Triangular meshes of the symmetric plane for laminar flow past a delta wing.

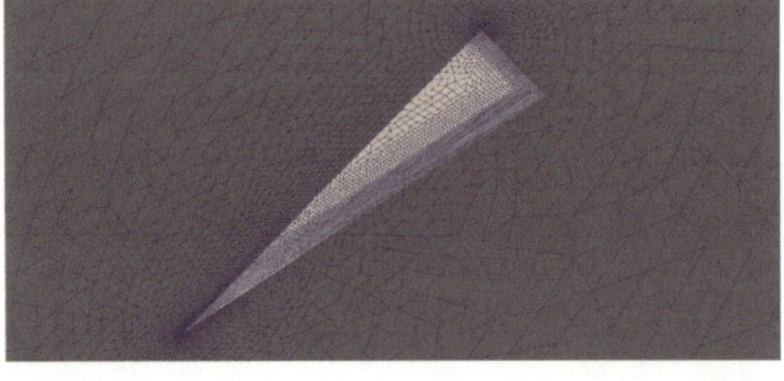

Figure 7.
Triangular meshes of delta wing surface.

Figure 8.
Vorticity magnitude contours and stream traces for delta wing.

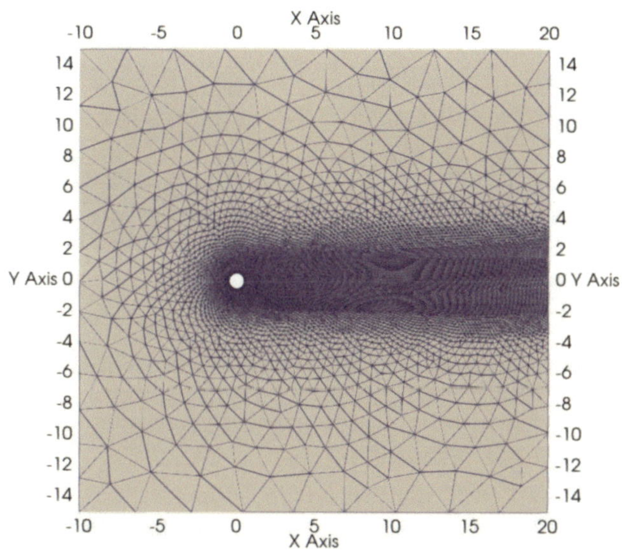

Figure 9.
Meshes for von Kármán vortex street.

5. Conclusions

High-order reconstructed discontinuous Galerkin methods based on first-order hyperbolic systems for Navier–Stokes equations have been developed and presented. The presented work is based on a new formulation for the hyperbolic Navier–Stokes system, namely HNS20G, which introduces the gradients of density, velocity, and temperature as additional variables. The numerical cases demonstrate that the presented method is compact, efficient, and robust.

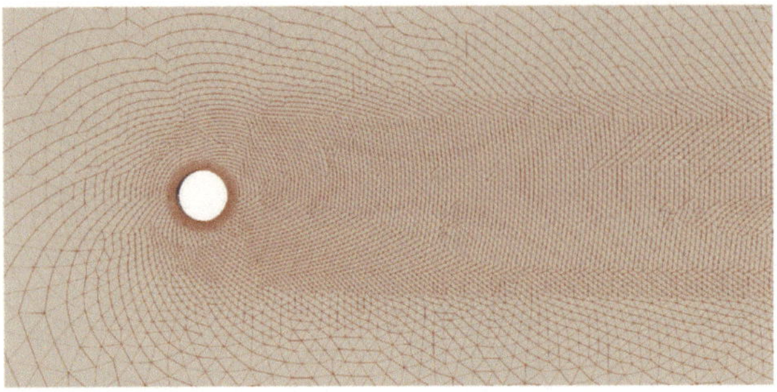

Figure 10.
Zoom-in meshes for von Kármán vortex street.

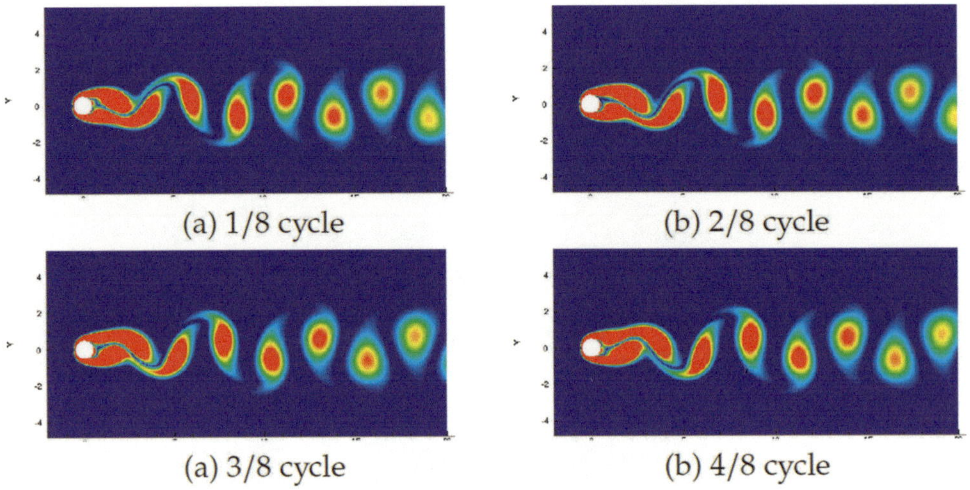

Figure 11.
Vorticity magnitude for von Kármán vortex street behind a cylinder at $Re_{\infty} = 100$ *within half cycle.*

Acknowledgements

The authors would like to express their sincere thanks to Prof. Hong Luo at North Carolina State University and Dr. Hiroaki Nishikawa at the National Institute of Aerospace, who has provided many useful insights and thoughts during the development.

Author details

Lingquan Li[1*†] and Jialin Lou[2†]

1 George Mason University, Fairfax, VA, Unitied States

2 Soochow University, Suzhou, China

*Address all correspondence to: lli30@gmu.edu

† These authors contributed equally.

References

[1] Peshkov I, Romenski E. A hyperbolic model for viscous Newtonian flows. Continuum Mechanics and Thermodynamics. 2016;**28**(1):85-104

[2] Dumbser M, Peshkov I, Romenski E, Zanotti O. High order ADER schemes for a unified first order hyperbolic formulation of continuum mechanics: Viscous heat-conducting fluids and elastic solids. Journal of Computational Physics. 2016;**314**:824-862

[3] Nishikawa H. First-, second-, and third-order finite-volume schemes for diffusion. Journal of Computational Physics. 2014;**256**:791-805

[4] Mazaheri A, Nishikawa H. Improved second-order hyperbolic residual-distribution scheme and its extension to third-order on arbitrary triangular grids. Journal of Computational Physics. 2015; **300**:455-491

[5] Lou J, Liu X, Luo H, Nishikawa H. Reconstructed discontinuous Galerkin methods for hyperbolic diffusion equations on unstructured grids. In: 55th AIAA Aerospace Sciences Meeting, AIAA Paper. Vol. 310. Reston, VA: American Institute of Aeronautics & Astronautics (AIAA); 2017

[6] Chamarthi AS, Nishikawa H, Komurasaki K. First order hyperbolic approach for anisotropic diffusion equation. Journal of Computational Physics. 2019;**396**:243-263

[7] Lou J, Li L, Luo H, Nishikawa H. Reconstructed discontinuous Galerkin methods for linear advection–diffusion equations based on first-order hyperbolic system. Journal of Computational Physics. 2018;**369**:103-124

[8] Nishikawa H, Liu Y. Hyperbolic Navier-stokes method for high-Reynolds-number boundary layer flows. In: 55th AIAA Aerospace Sciences Meeting. Reston, VA: American Institute of Aeronautics & Astronautics (AIAA); 2017. p. 0081

[9] Anderson JD, Wendt J. Computational Fluid Dynamics. Vol. 206. New York: McGraw-Hill; 1995

[10] Toro EF. Riemann Solvers and Numerical Methods for Fluid Dynamics: A Practical Introduction. Berlin/ Heidelberg, Germany: Springer-Verlag; 2013

[11] Löhner R. Applied Computational Fluid Dynamics Techniques: An Introduction Based on Finite Element Methods. England: John Wiley & Sons; 2008

[12] Li L, Lou J, Luo H, Nishikawa H. A new formulation of hyperbolic Navier-stokes solver based on finite volume method on arbitrary grids. In: 2018 Fluid Dynamics Conference. Reston, VA: American Institute of Aeronautics & Astronautics (AIAA); 2018. p. 4160

[13] Reddy JN. Introduction to the Finite Element Method. New York City, NY: McGraw-Hill Education; 2019

[14] Bassi F, Rebay S. A high-order accurate discontinuous finite element method for the numerical solution of the compressible Navier–stokes equations. Journal of Computational Physics. 1997; **131**(2):267-279

[15] Bassi F, Crivellini A, Rebay S, Savini M. Discontinuous Galerkin solution of the Reynolds-averaged Navier–stokes and k–? turbulence model equations. Computers & Fluids. 2005;**34** (4–5):507-540

[16] Kannan R, Wang ZJ. The direct discontinuous Galerkin (DDG) viscous flux scheme for the high order spectral volume method. Computers & Fluids. 2010;**39**(10):2007-2021

[17] Cheng J, Yang X, Liu X, Liu T, Luo H. A direct discontinuous Galerkin method for the compressible Navier–stokes equations on arbitrary grids. Journal of Computational Physics. 2016; **327**:484-502

[18] Nishikawa H. A first-order system approach for diffusion equation. I: Second-order residual-distribution schemes. Journal of Computational Physics. 2007;**227**(1):315-352

[19] Nishikawa H. New-generation hyperbolic Navier-stokes schemes: O (1/h) speed-up and accurate viscous/heat fluxes. In: 20th AIAA Computational Fluid Dynamics Conference. Reston, VA: American Institute of Aeronautics & Astronautics (AIAA); 2011. p. 3043

[20] Nishikawa H. Alternative formulations for first-, second-, and third-order hyperbolic Navier-stokes schemes. In: 22nd AIAA Computational Fluid Dynamics Conference. Reston, VA: American Institute of Aeronautics & Astronautics (AIAA); 2015. p. 2451

[21] Li L, Lou J, Nishikawa H, Luo H. Reconstructed discontinuous Galerkin methods for compressible flows based on a new hyperbolic Navier-stokes system. Journal of Computational Physics. 2021; **427**:110058

[22] Johnson TA. Numerical and Experimental Investigation of Flow Past a Sphere up to a Reynolds Number of 300. Iowa City, Iowa: The University of Iowa; 1996

Section 2

Application of Computational Fluid Dynamics

Chapter 3

Computational Fluid Dynamics for Advanced Characterisation of Bioreactors Used in the Biopharmaceutical Industry – Part I: Literature Review

Stefan Seidel, Cedric Schirmer, Rüdiger W. Maschke, Lia Rossi, Regine Eibl and Dieter Eibl

Abstract

Computational fluid dynamics (CFD) is a widely used tool for investigating fluid flows in bioreactors. It has been used in the biopharmaceutical industry for years and has established itself as an important tool for process engineering characterisation. As a result, CFD simulations are increasingly being used to complement classical process engineering investigations in the laboratory with spatially and temporally resolved results, or even replace them when laboratory investigations are not possible. Parameters that can be determined include the specific power input, Kolmogorov length, hydrodynamic stress, mixing time, oxygen transfer rate, and for cultivations with microcarriers, the N_{S1} criterion. In the first part of this series, a literature review illustrates how these parameters can be determined using CFD and how they can be validated experimentally. In addition, an overview of the hardware and software typically used for bioreactor characterisation will also be provided, including process engineering parameter investigations from the literature. In the second part of this series, the authors' research results will be used to show how the process engineering characterisation of mechanically driven bioreactors for the biopharmaceutical industry (stirred, orbitally shaken, and wave-mixed) can be determined and validated using CFD.

Keywords: mixing time, oxygen mass transfer, power input, process engineering characterisation, single-use bioreactor, validation

1. Introduction

Bioreactors are closed systems that allow the cultivation of microbial, animal, and human cells in a controlled, aseptic environment. The simplest bioreactors are not instrumentalised (e.g., shake flasks) and dominate in mL-scale process development. In contrast, bench-top (10 L working volume), pilot (≥50 L working volume), and

IntechOpen

industrial-scale (working volumes of cubic metres) bioreactors are typically equipped with sensors, which allow regulation and adjustment of the cultivation process. At this time, the primary goal of such cultivation is the production of inocula, and bio-, cell-, or gene therapeutics. The bioreactor type and its operation parameters influence the growth and production behaviour of the production organism used. For example, the often-used microorganism *Eschericha coli* has high oxygen demands [1, 2] and a need for short mixing times [3] due to its fast metabolism. In addition, microorganisms such as *E. coli* are extremely robust thanks to their cell walls and small size. Therefore, the bioreactor can be operated with a high specific power input. Biopharmaceutical production with *E. coli* has a long tradition and dates back to 1982 when the production of insulin was approved by the Food and Drug Administration in the USA [4].

Today, most biopharmaceuticals, such as monoclonal antibodies, therapeutic hormones and many vaccines, are produced using mammalian cells, with Chinese hamster ovary (CHO) cells being the most frequently used [5]. Furthermore, human embryonic kidney (HEK) cells, Madin-Darby canine kidney (MDCK) cells, murine myeloma cells (NS0), and baby hamster kidney (BHK) cells are used [6]. The advantage of mammalian cells is that they can produce more complex molecules and can also perform glycosylation, which makes them even more attractive to use [7]. Furthermore, mammalian cells generally do not require as much oxygen as *E. coli* [1, 8]. However, they are larger and lack a cell wall, which makes them much more sensitive to fluid dynamic stress, a factor that has to be taken into account when selecting the bioreactor system and its operation parameters [9, 10].

In addition to mammalian cells and microorganisms, plant cells, insect cells and, more recently, stem cells are also used in biotechnological manufacturing processes. These cells are either continuous cells (unlimited life span) or primary cells (limited life span) and grow adherently (they need a surface to attach to, such as planar plastic surface, membranes or microcarriers) or in suspension.

Bioreactors can be classified in different ways. If the reusability of the cultivation container is taken into account, a classification into single-use and reusable systems can be made [11, 12]. If the type of mass transfer is taken into account, they can be categorised into static and dynamic systems, the latter of which can be further divided into mechanically, pneumatically and hydraulically driven bioreactors, depending on the type of power input [13, 14]. Mechanically driven systems, which are either stirred, orbitally shaken, or wave-mixed, predominate [13], the last of which are exclusively available as single-use variants. Single-use bioreactors have become very popular over the past 20 years [15]. The cultivation containers (rigid plastic vessels or flexible bags) are pre-sterilised, can be used immediately, and are disposed of after the cultivation has been completed. This makes the cultivation safer, helps to save time, may reduce costs, and often contributes to a lower impact on the environment despite the plastic waste that is generated. Efforts by manufacturers of single-use bioreactors have helped make them more user-friendly, for example, by replacing weak points such as the film material of the first and second bioreactor generation with improved ones that are less problematic in terms of leachables and extractables [16]. Currently, single-use bioreactors up to 6 m^3 working volume are available on the market [15].

Stirred bioreactors (**Figure 1a**) are characterised by a mostly cylindrical vessel containing a stirring shaft with one or more stirrers. The dimensioning of the vessel as well as the type, number and positioning of the stirrer(s), which introduce power into the system, differ depending on the application. The oxygen required by the cells is added via the fluid surface or by active gassing with a sparger, depending on the application. Stirred bioreactors range from the mL-scale in research and development

Figure 1.
Computer-generated images (CGI) of common mechanically driven bioreactor systems. a) Stirred bioreactor where the power is brought into the system by rotation of the stirring shaft and the stirrer (rotation marked by green arrow). b) Shake flask, which is an orbitally shaken bioreactor. Here the power is brought into the system by the orbitally shaking of the platform (movement of the platform marked by green arrow). c) Wave-mixed bioreactor, in which the power is brought in by the back and forth tilting of the bioreactor (rotation around tilting axis marked by green arrow).

to more than 100 m^3 in production processes [12]. In orbitally shaken systems, such as the shake flasks (**Figure 1b**), power is introduced into the system through the walls of the vessel by moving the entire system orbitally on a platform, and oxygen is typically only introduced via the free surface. Wave-mixed systems (**Figure 1c**) are also surface-aerated however, in these systems power is introduced by tilting the cultivation bag to create one or two degrees of freedom (DOF) of motion, resulting in an undulation in the fluid inside the bioreactor.

Bioreactors should be both, technically characterised and biologically qualified. Details on biological qualification of bioreactors are described by Schirmer et al. [17, 18]. Technical or so-called classical process engineering characterisation includes the determination of specific power input, mixing or residence time, and oxygen transfer [19]. If adherent cells such as mesenchymal and induced pluripotent stem cells are grown on microcarriers, investigation of the suspension behaviour can also be beneficial [20]. Characterisation is typically performed in the laboratory by means of experiments (Section 4). However, such characterisation is resource-intensive (financial and labour costs). Moreover, only bioreactor systems that already physically exist can be characterised. However, with the help of computational fluid dynamics (CFD), it is possible to eliminate some of these disadvantages, making it is possible to characterise and optimise digital systems prior to construction. This allows expensive production stops to be avoided, and more complex investigations to be performed. Nevertheless, there are certain disadvantages associated with the use of CFD. Well-trained personnel and a lot of computing power are required. Validation is necessary to obtain reliable results, since CFD is only a model of the real world that is limited by a finite degree of accuracy. In this chapter, process parameters and their determination by CFD are described and discussed. In addition, possible hardware and software solutions are also presented as well as validation methods.

2. Process engineering characterisation

The specific power input P/V is one of the central process engineering parameters. It describes how much power per volume is introduced into the bioreactor system. The power input influences various other parameters such as oxygen transfer, mixing intensity and fluid dynamic stress [21]. As already mentioned, a distinction can be made between mechanical, pneumatic, and hydraulic power inputs. Typical specific power inputs for microbial cultivations are >5 kW m^{-3} [22] and thus significantly higher than for mammalian cells (5 to 310 W m^{-3}) [23, 24]. The specific power input, which is mechanically introduced into the bioreactor system can be determined by the torque M at a given stirrer speed n for volume V (eq. (1)).

$$P/V = \frac{2 \cdot \pi \cdot n \cdot M}{V} \tag{1}$$

The specific power input determined by this method corresponds to the average power input. The average power input is equal to the product of the average energy dissipation rate $\overline{\varepsilon}$ and the density ρ (eq. (2)) [9]. However, the energy dissipation rate in a bioreactor is not uniform and can differ by several orders of magnitude. For example, in a stirred bioreactor, the energy dissipation rate is maximum at the stirrer tips and decreases sharply as the distance from the tips increases. These differences in energy dissipation rate can be expressed in terms of hydrodynamic heterogeneity, which is the quotient of ε_{max} and $\overline{\varepsilon}$. Orbitally shaken systems are characterised by high homogeneity, where $\varepsilon_{max}/\overline{\varepsilon}$ ranges from 1 to 18 [25, 26], and wave-mixed bioreactors lie between orbitally shaken and stirred bioreactors (8.8 to 32.0) [27]. The greatest heterogeneity can be observed in stirred bioreactors, with values of up to 66 for

6-blade pitched turbines and as high as 147 for Rushton turbines [28]. Kolmogorov's microscale theory, which is often used as an indicator of possible cell damage, can be derived directly from the turbulent energy dissipation rate. Assuming isotropic turbulence, this theory states that eddies break down into smaller and smaller eddies and dissipate into heat at the Kolmogorov length λ_k. Eddies that are significantly larger than cells (e.g., 20 μm for CHO cells) damage the cells less and simply transport the cells with them. However, if the vortices are in the size range of the cells or smaller, this can lead to damage to the cell membrane. In addition to the Kolmogorov length, fluid dynamic stress can also be used to assess possible cell damage. This is composed of normal σ and shear τ stresses. Lethal and sublethal hydrodynamic stress varies greatly from one organism to another and can even vary for the same cell type [29]. In addition to the cell type, culture media composition [30, 31] and exposure duration also play crucial roles.

$$P/V = \frac{\sum(\varepsilon_i)V_i\rho}{V} = \overline{\varepsilon} \cdot \rho \tag{2}$$

The mixing time is another process engineering parameter which is of importance for large bioreactors. The mixing time corresponds to the time required to achieve a defined mixing quality M (typically 95% in biotechnology, eq. (3)). The mixing time should be kept as low as possible so that no oxygen and nutrient limitations or pH fluctuations occur. However, few exact values are published in the literature. For example, Anane et al. [32] showed that a clear metabolic switch occurred in a CHO cell line when the mixing time was longer than 90 s. However, the maximum mixing time suggested by Löffelholz et al. [33] for litre-scale devices is 30 s, which is below this critical value. A detailed description of the decolourisation method, which can be used to determine the mixing time, can be found in Bauer et al. [19] or in Maschke et al. [10].

$$M(t) = 1 - \frac{|c(t) - c_\infty|}{c_\infty} \tag{3}$$

A sufficient supply of oxygen is of critical importance in aerobic cultivations. In contrast to other nutrients, oxygen must be continuously added to the system, since it does not dissolve well in water-like media (Newtonian culture broths). The amount of oxygen required for a cultivation can be determined by the oxygen uptake rate (OUR), which is the product of the cell density and the specific oxygen uptake rate q_{O_2}. An overview of typical specific oxygen uptake rates can be found in Maschke et al. [10] and Seidel et al. [2]. To ensure sufficient oxygen supply, the oxygen transfer rate (OTR) must be equal to or greater than the OUR. The OTR is the product of the volumetric oxygen mass transfer coefficient k_La and the difference between the oxygen saturation concentration C^* and the dissolved oxygen concentration in the bulk liquid C.

For process characterisation purposes, the k_La value is of particular interest. This value is the product of the liquid side mass transfer coefficient k_L and the specific interface a. In practice, the k_La value is typically determined directly. The most common methods are the gassing-out method, sulfite method and the respiratory gassing-out method, the latter of which is a biotic method. A detailed description of methods for determining the k_La value can be found in Seidel et al. [2] or Bauer et al. [19]. However, there are also measurement methods that allow the determination of just the specific interface (Section 4).

Another important parameter is the suspension criterion N_{S1}. The characterisation of the sedimentation behaviour is particularly important for the cultivation of cells growing adherently on microcarriers, such as stem cells, which are also sensitive to hydrodynamic stress. Various authors have used the N_{S1} criterion to successfully grow stem cells on microcarriers [34–36]. This criterion describes the stirrer speed and thus the minimum power input required to completely suspend all microcarriers. It should be noted that at this speed only a suspension of the microcarriers is guaranteed, but not a homogeneous distribution in the cultivation vessel. This criterion ensures that not only are the cells subjected to the minimum necessary hydrodynamic stress, but also that there is no nutrient limitation due to heap formation. The N_{S1} criterion can be determined using a camera without the need for extensive apparatus.

3. CFD models

CFD is used in various fields, including biotechnology, for research, development and education [37]. The most widely used method is the finite volume method that discretises the Navier–Stokes equations. In addition, this method, which is used in most commercial and open source CFD codes (e.g., Ansys Fluent and CFX, Siemens Simcenter Star-CCM+, OpenFOAM), the finite element method (Autodesk CFD) or the Lattice-Boltzmann method (M-Star CFD, OpenLB) are also used. In the latter, the Boltzmann equation is discretised instead of the Navier–Stokes equation [38]. In this chapter, only the finite volume method will be discussed. To describe a single-phase laminar flow in a bioreactor, the conservation equations for mass and momentum are discretised and solved iteratively. Single-phase considerations are only suitable for stirred bioreactors where a constant fluid surface can be assumed (no vortex formation).

For most biotechnological applications, a turbulent flow needs to be generated to ensure sufficient mixing for the cells to be suspended, for the oxygen to be dispersed and for the nutrients to be evenly distributed. This is, however, not the case for bioreactors used for cultivating cells that may be sensitive to hydrodynamic stress. Several methods exist to account for turbulent flow, with the Reynolds-averaged Navier–Stokes (RANS) approach being the most widely used due to its economic advantages. This approach fully describes the turbulence using an additional model and only calculates the average flux. There are a number of different RANS turbulence models. However eddy-viscosity-based models, such as the k-ε-model (Eqs. (4) and (5)) and k-ω-model (equations can be found in Wilcox [39]), are the most widely used [40, 41]. If the phenomenon of turbulence needs to be considered in more detail, the unsteady-RANS, Hybrid RANS large eddy simulation (LES), LES or direct numerical simulation (DNS) can be used with increasing computational complexity. In fact, DNS is so computationally intensive that it is only used for simple geometries in research [42]. A detailed overview of turbulence models is described in Wilcox [39] and Rodriguez [43].

$$\frac{\partial(\rho k)}{\partial t}+\frac{\partial}{\partial x_i}(\rho u_i k)=\frac{\partial}{\partial x_i}+\left(\mu+\frac{\mu_t}{\sigma_k}\right)\frac{\partial k}{\partial x_i}+\mu_t\left(\frac{\partial u_i}{\partial u_j}+\frac{\partial u_j}{\partial u_i}\right)\frac{\partial u_j}{\partial x_i}-\rho\varepsilon \tag{4}$$

$$\frac{\partial(\rho \varepsilon)}{\partial t}+\frac{\partial}{\partial x_i}(\rho u_i \varepsilon)=\frac{\partial}{\partial x_i}+\left(\mu+\frac{\mu_t}{\sigma_\varepsilon}\right)\frac{\partial \varepsilon}{\partial x_i}+C_1\frac{\varepsilon}{k}\mu_t\left(\frac{\partial u_i}{\partial u_j}+\frac{\partial u_j}{\partial u_i}\right)\frac{\partial u_j}{\partial x_i}+C_2\rho\frac{\varepsilon^2}{k} \tag{5}$$

Mechanically driven bioreactors are characterised according to the method used to introduce power into the system, described in **Figure 1**. For stirred systems, two different methods are mainly used to model the motion of the stirrer. With the frozen rotor method (also known as multiple reference frame method) the stirrer is not rotated, but in the zone around the stirrer the conservation equations are solved in a rotating reference system. Thus, there is no displacement or recalculation of the computational mesh, which makes this method resource efficient. However, this method is only suitable for steady-state considerations of stirred bioreactors. Another method is the sliding mesh approach, in which the mesh of the stirrer and stirred zone is shifted relative to the rest of the bioreactor. This method is significantly more computationally intensive than the frozen rotor method, but allows for a transient approach. There are also two different methods for orbitally shaken and wave-mixed bioreactors, both of which use a transient approach. The computational mesh can be manipulated by rotation and/or translation to mimic motion [44]. A less common method is to manipulate only the direction of the forces acting within the bioreactor. For example, Zhu et al. [45] successfully modelled orbitally shaken systems using manipulation of centrifugal forces.

In terms of the process engineering characterisation, a single-phase simulation can be used to determine the flow field, as well as the non-aerated specific power input, with the power input being determined analogously to eq. (1). Instead of measuring the torque, it can be determined using eq. (6) [46], where F_p and F_τ correspond to the pressure and viscous force, and r to the distance from the mesh cell centre to the axis of rotation (Eqs. (7) and (8)).

$$M = \sum_i \left(r_i \times F_{p_i} + r_i \times F_{\tau_i}\right) \tag{6}$$

$$F_{p_i} = p_i \cdot A_i \tag{7}$$

$$F_{\tau_i} = \mu \cdot \frac{\partial u_i}{\partial x_i} \cdot A_i \tag{8}$$

Normal and shear gradients can be determined from the velocity field or its gradients. A detailed derivation is described in Wollny [47]. By using RANS turbulence models, the turbulent energy dissipation rate ε can be determined. This is either calculated directly, for example, when using the k-ε-model, or can be determined from k and ω when using the k-ω-model (eq. (9)). The energy dissipation rate can be used to predict the Kolmogorov length λ_k (eq. (10)). If this is determined for each mesh cell, a Kolmogorov length distribution in the bioreactor can be determined [44].

$$\varepsilon = k \cdot \omega \cdot \beta^* \tag{9}$$

$$\lambda_k = \left(\frac{\nu^3}{\varepsilon}\right)^{\frac{1}{4}} \tag{10}$$

The mixing time in stirred bioreactors can also be determined based on a stationary single-phase simulation. For this purpose, the converged stationary solution of the velocity field is used as a basis. Subsequently, a virtual tracer T_m can be fed into the system via a scalar transport equation (eq. (11)) [48].

$$\frac{\partial T_\mathrm{m}}{\partial t} + \nabla \cdot (V \cdot \mathrm{Tm}) - \nabla^2((D + D_\mathrm{t}) \cdot T_\mathrm{m}) = 0 \tag{11}$$

The subsequent transient observation shows how the tracer is distributed in the system by convection and diffusion (including turbulent diffusion D_t) over time. Analogously to the decolourisation method, the mixing time can thus be determined.

If the fluid flow in orbitally shaken, wave-mixed or vortex-forming stirred bioreactor systems is being investigated, the gas phase has to be modelled in addition to the liquid phase. Different methods exist for the investigation of muliphase systems. However, the volume of fluid (VOF) method is suitable for continuous two-phase systems, in which the continuity can be assumed. The continuity equation corresponds to eq. (12) and the momentum equation to eq. (13). The VOF method uses a mixed fluid in which the physical properties χ such as density ρ and viscosity μ are weighted by the phase fraction α (eqs. (14) and (15)). The surface tension σ can be considered here by using, for example, the continuum surface force (CSF) model [49].

$$\nabla \cdot \vec{v} = 0 \tag{12}$$

$$\frac{\partial \rho \vec{v}}{\partial t} + \nabla \cdot \left(\rho \vec{v} \vec{v}\right) = -\nabla p + \rho \vec{g} + \nabla \cdot \nu \left(\nabla \vec{v} + \left(\nabla \vec{v}\right)^T\right) + \sigma \kappa \nabla \alpha_{\text{liquid}} \tag{13}$$

$$\chi = \sum \chi_i \alpha_i, \quad \chi \in [\rho, \nu] \tag{14}$$

$$\sum \alpha_i = 1 \quad \forall \alpha_i, \{\alpha_i | 0 \leq \alpha_i \leq 1\} \tag{15}$$

Similarly to the single-phase simulation, the power input can also be determined using the torque value. Here, the phase fraction must be taken into account as well as the fact that this is a transient determination, which may have to be averaged over time for the average specific power input. Shear and normal gradients as well as Kolmogorov lengths can also all be determined in a similar way to the single-phase simulations.

The $k_L a$ value can also be determined for these systems, which employ pure surface-aeration. In contrast to experimental investigations, k_L and a are determined individually in CFD simulations. For surface-aerated systems, the specific interface can be determined from the quotient of the liquid surface area and the working volume, with the liquid surface area determined using an iso-surface with $\alpha_{\text{water}} = 0.5$. The k_L value can be approximated by several models. Higbie's [50] model is widely used (eq. (16)) and requires the energy dissipation rate to be determined. The mixing time can be determined by a virtual tracer. However, in VOF models it is important to ensure that the virtual tracer does not migrate into the gas phase. There are different model approaches used in commercial and open source CFD codes (phasescalar-transport in OpenFOAM, user-defined scalar in Ansys Fluent).

If, however, a dispersed system is to be considered, the VOF model is not able to represent this complex phenomena. Dispersed systems include active aerated systems and systems with particles such as microcarriers. In contrast to the VOF model, the Euler-Euler model does not use a mixed fluid, but instead solves the conservation equations for each phase individually. However, only one pressure field is calculated for all phases. Nevertheless, with the Euler-Euler model, interfacial forces such as drag, lift and virtual mass force, can also be taken into account, which makes it possible to model particles of a defined size (different particle sizes can be modelled by additional phases, but this is a computationally intensive method). These forces and their computations are described in detail in Seidel et al. [2], Pourtousi et al. [51] and Chuang and Hibiki [52]. If a known constant bubble diameter d can be assumed, the specific interface a can also be determined with the Euler-Euler model (eq. (17), for mono-disperse systems: $d_{32} \mathrel{\hat{=}} d$) [53].

$$k_L = \frac{2}{\sqrt{\pi}} \varepsilon \nu \left(\frac{D_{O_2}}{\nu}\right)^{\frac{1}{4}} \tag{16}$$

$$a = \frac{A_g}{V} = \frac{6\alpha}{d_{32}} \tag{17}$$

Seidel and Eibl [8] have been able to show that in a 3 L stirred bioreactor, under the same simulation conditions, the k_La value can be described as a power function of the bubble diameter with a negative exponent. However, combining CFD with population balance modelling (PBM) can minimise the heavy dependence of the simulated k_L value on the selected bubble diameter. This is achieved by modelling a population of bubbles with a bubble size distribution. In addition, PBM can also be used to model complex phenomena such as bubble breakup and coalescence by describing the change in the number density function of the gas bubble diameters, with the source term consisting of four components: death by coalescence, birth by coalescence, death by breakup, birth by breakup. A detailed description and overview of closure models for the population balance equations can be found in Liao and Lucas [54, 55], and a description of how the choice of coalescence and breakup models affects the k_La value is provided by Seidel and Eibl [8]. Among the most frequently used approaches are the class method, which is used in OpenFOAM, and the method of moments, which is used in Ansys Fluent. A detailed overview of possible methods can be found in Nguyen et al. [56] and Li et al. [57], and their advantages and disadvantages are discussed in Seidel et al. [2].

The Euler-Euler model can also be used to perform suspension investigations and to determine the N_{S1} criterion [58–60]. However, CFD investigations concerning the N_{S1} criterion are not very widespread. In order to model the microcarriers as a solid phase, a Granular Flow Model (GFM) is required, with the kinetic theory of granular flow (KTGF) approach being widely used for this purpose [61, 62]. For modelling purposes, a granular pressure is introduced, which requires a gradient term for the granular phase momentum equation. In order to calculate the granular pressure, it is necessary to also calculate the granular temperature, which influences the collision behaviour of the particles [63]. An example of how the granular pressure can be modelled can be found in Syamlal et al. [64]. When using the Euler-Euler model, instead of modelling the individual particles (gas bubbles or microcarriers), only their phase fractions are modelled. If, however, the individual particles need to be modelled, the Euler-Lagrange model should be used, which describes the paths of the particles using an ordinary differential equation [65]. However, the Euler-Lagrange model does not consider the influence of turbulent flows on the particle path, which can be accounted for using the discrete random walk model [66]. It is a fact that the Euler-Lagrange model is used less frequently for biotechnological applications than the Euler-Euler model [2]. The main reason for this is that the computation time increases linearly as particle numbers increase [67]. Zieringer and Takors [68] recommend a maximum particle fraction of 10% for the use of the Euler-Lagrange model. Accordingly, using 10 $g\ L^{-1}$ (1% particle fraction) microcarriers, as Jossen et al. [69] did, would require approximately $40 \cdot 10^6$ carrier L^{-1} to be individually modelled [70].

Depending on the bioreactor system and the process parameters being investigated, different models should be considered in order to avoid uneconomical CFD simulations. It should be noted that irrespective of the model used for a CFD simulation, it will never be a true representation of reality. In addition to modelling errors, discretisation errors, iteration-convergence errors, rounding errors, programming errors and user errors can all lead to inaccuracies. Rounding errors are typically negligible (double-precision floating point numbers) and user errors should ideally be entirely avoidable. Discretisation error arise from the spatial discretisation of the mesh cells and the temporal discretisation of the time step sizes. A widely used approach for estimating discretisation errors arising from the choice of mesh is the Richardson extrapolation or the associated grid convergence index [71–73], which considers the change in the target parameter (e.g. specific power input or mixing time) using meshes of different resolutions. A detailed description of this procedure is described in Ramírez et al. [74] and Seidel et al. [44]. Alternatively, methods like the grid systematic refinement or the curve fitting method can also be used [75].

If the estimated discretisation error is known, an economically acceptable mesh can be chosen, since the computation time increases exponentially as the number of mesh cells increases, while the discretisation error approaches zero asymptotically [76]. For transient simulations, both spatial and temporal discretisation play a role. If the chosen time step size is too long, this can lead to instability and inaccuracies in the simulation. Like finer computational meshes, computing time also increases linearly as the time steps becomes shorter. The Courant-Friedrichs-Lewy (CFL) number helps to estimate the time step size [77]. If an explicit time integration scheme is used, CFL values above 1 can lead to non-physical values and divergence. A CFL number below 1 is preferable even in implicit schemes in order to avoid large errors. The CFL number, which is calculated individually for each mesh cell, can vary significantly within a single bioreactor model. Therefore, the maximum CFL number of any single mesh cell should not exceed 1 [78]. In order to determine the total error of the CFD results, validation by means of physical experiments is necessary. The most common validation procedures for CFD simulations of bioreactors are described in section 4.

4. Validation

The validation of CFD simulations is of great importance and is defined by the American Institute of Aeronautics and Astronautics [79] as "The process of determining the degree to which a model is an accurate representation of the real world from the perspective of the intended uses of the model." In order to perform a validation, different experimental investigations can be carried out. It should be noted that the experiments are also not free of measurement errors, which must also be taken into account in the statistical statement. Depending on the target process variable, different methods are suitable for validating different bioreactor simulations. One of the most fundamental factors is the flow pattern. Particle image velocimetry (PIV) [80–82], laser doppler anemometry [83, 84] and laser induced fluorescence [81, 85] are suitable for measuring the velocities in a bioreactor and are especially used for stirred bioreactors. For orbitally shaken [86, 87] and wave-mixed [88] systems, often only the liquid height or distribution is measured.

If the power input is of relevance, it is reasonable to use it as a validation criterion. Thus, the measurement of the torque around the rotation axis is suitable for stirred [74, 89], orbitally shaken [90, 91] and wave-mixed [92] bioreactors. Alternatively, the specific power input can be determined either by measuring the electric current supplied to the motor [93, 94] or calorimetrically [95, 96]. Villiger et al. [97] described a method, which allows the maximum hydrodynamic stress to be determined using poly(methyl methacrylate) aggregates [98], which is also suitable for validation purposes. In addition, the mixing time can also be used as a validation parameter.

Depending on whether the mixing time is evaluated globally or locally in the CFD simulation, the decolourisation method or the conductivity measurement according to DECHEMA e.V. Working Group for Single-Use Technology are both suitable for validation [19]. However, these are not the only valid approaches, for example, Vivek et al. [99] used a Raman spectroscopy probe to experimentally determine the mixing time to validate CFD-determined mixing times.

For aerated systems, either the $k_{L}a$ value can be used directly as a validation parameter, or the bubble size distribution can be used. The $k_{L}a$ value can be measured using different methods described in Seidel and Eibl [8] or Garcia-Ochoa and Gomez [100], and the bubble size distrbution by shadowgraphy [101], optical multimode online probe [102], capillary suction probe technique [103, 104], SOPAT endoscopy [105, 106] or focused beam reflectance [107, 108] measurements. The gas holdup, which is typically determined optically, by the differential pressure method [109] or radar level gauge [93] is also suitable for aerated systems.

For suspension studies with microcarriers or other particles, it is advisable to use the suspension behaviour or the N_{S1} criterion for validation. For this purpose, the particles that are to be suspended are added to an unstirred bioreactor. Then the speed is increased stepwise until the N_{S1} criterion is achieved. The evaluation is either performed by eye [110] or recorded with a camera [59, 111–113]. Delafosse et al. [113] and Loubière et al. [59] have refined this evaluation method by measuring the light attenuation using a light source mounted on one side of the bioreactor and the camera on the opposite side. Most authors have recorded lateral images as well as images from below, using a mirror for the latter. For the validation of their CFD simulations, Zhang et al. [114] recorded images with a high-speed camera. However, they did not examine cells grown on microcarriers, but rather immobilised *Lactobacillus* cells.

5. Hardware and software

In order to obtain a final result, several steps are necessary, which can be divided into preprocessing, processing and postprocessing [76, 115]. The individual steps are explained in more detail in **Figure 2**. Typically, several of these steps are included in today's CFD software solutions. Using the Ansys software solution, the geometry can be created (SpaceClaim), the mesh can be generated (Meshing), the calculations can be performed (Fluent) and with CFD-Post the postprocessing can be completed. However, separate software is often used for geometry generation, such as Autodesk Inventor, SolidWorks, Solid Edge or Salome, and for postprocessing steps such as Tecplot and Paraview. The most widely used CFD software solutions for bioreactor modelling are Ansys Fluent and CFX (commercial) and OpenFOAM (open source). However, Simcenter Star-CCM+ from Siemens, Autodesk CFD, COMSOL Multiphysics and M-Star CFD are also used, with the latter specifically advertised for bioreactor applications (all commercial software).

In addition to the software, the hardware is also of critical importance for a simulation to be economic [116], with the hardware performance, the purchase price and the power consumption all playing a role. All current CFD software solutions allow for the parallelisation of the calculations. For this purpose, the computational mesh is partitioned into different domains. Individual processors then execute the computations in the individual domains. The communication between the domains is regulated, as usual in parallel computing, by means of the message passing interface (MPI) or other interfaces [116]. Different algorithms exist for creating partitions,

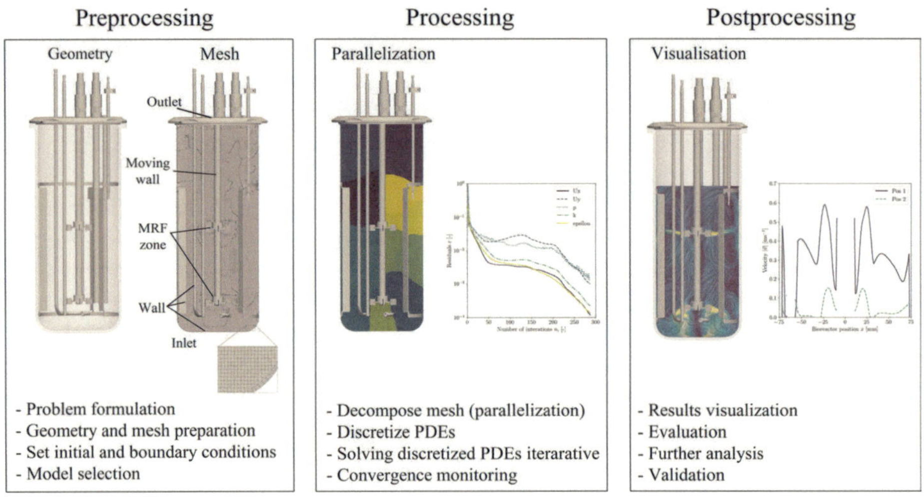

Figure 2.
Visualisation of the process steps ranging from problem definition to validation. The Minifors 2 stirred bioreactor from Infors AG hase been used as an example.

which differ in their degree of automation and partitioning time [117]. For example, the Scotch algorithm of Pellegrini and Roman [118], which is based on dual recursive bi-partitioning and implemented in OpenFOAM, only needs the number of domains to perform the process. However, this algorithm requires a longer partitioning time than less automated algorithms [117].

While parallelisation reduces the required computing time, higher degrees of parallelisation also require more data to be exchanged between the processors, which has a negative impact on computing time. **Figure 3** shows how the relative simulation time for modelling a stirred 3 L bioreactor changes as parallelisation increases (interFOAM, OpenFOAM v9). With the hardware setup described in Seidel and Eibl [8], the relative simulation time can be described as a power function with an exponent of -0.817. If there were no loss due to communication, an exponent of -1 would be expected. Harasek et al. [119] also performed parallelisation studies with up to 1024 cores, form which an exponent of -0.93 could be determined. According to Haddadi et al. [116], parallelisation should be performed in such a way that there are between 50,000 and 100,000 cells per domain, whereby the number of cells should be reduced as the complexity of the model increases.

Of the publications listed in **Table 1**, only one-third of the authors made statements about the hardware used. Between 4 [120] and 504 [121] cores were used for the calculations. Only five of the authors stated that they had used an HPC system [44, 121–124], with the remaining authors having used desktop machines. The current versions of Ansys Fluent and Siemens Simcenter Star-CCM+ emphasised the use of graphics processing units (GPU) instead of central processing units for the calculations. Multiple GPUs can also be used here. Benchmarks from Ansys and Siemens show that using GPUs can shorten simulation time at the same time as reducing purchasing costs and power consumption [125, 126]. Of the authors listed in **Table 1**, only authors who used M-Star CFD stated that a GPU was used (M-Star CFD relies exclusively on GPUs). COMSOL, Autodesk CFD, Ansys CFX and OpenFOAM are not able to use GPUs for calculation by default. There are, however, a number of modifications such as MixIT, that allow OpenFOAM to us GPUs [127, 128]. Current

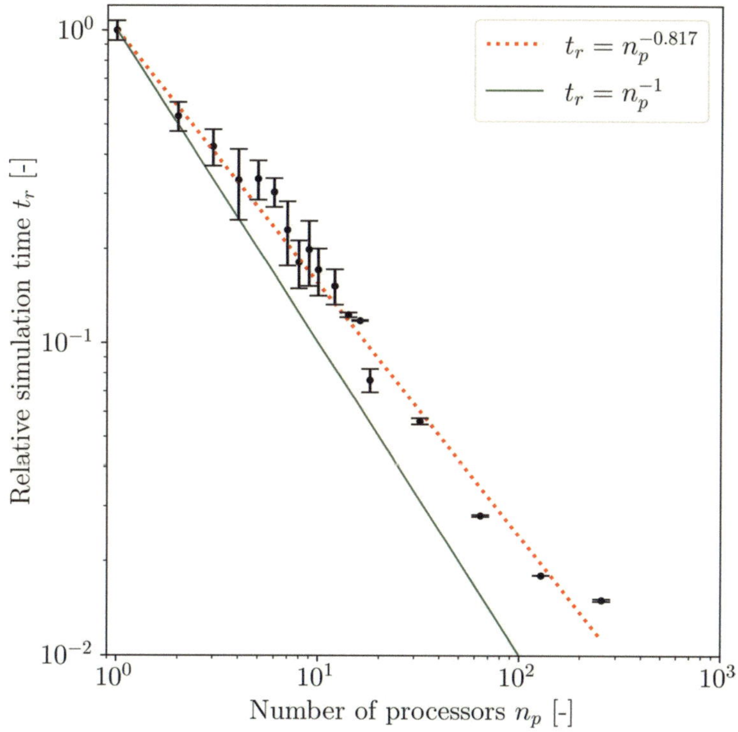

Figure 3.
Relative simulation time depending on the processors used. A stirred bioreactor, which was examined with OpenFOAM v9, was used as a test case (VOF model, transient observation for 10 s). All simulations were performed three times to capture temporal variance. In each case, the Scotch algorithm was used for decomposition. The hardware setup is described in more detail in Seidel and Eibl [8].

developments show that classical CFD simulations could be complemented or replaced by other techniques. Machine learning techniques can be used to accelerate simulations or to improve turbulence modelling [129], and physics-informed neural networks (PINN) are increasingly used, for their ability to perform calculations 200 times faster to the same degree of accuracy [130, 131]. Another technique, which currently has no real application for bioreactor modelling, is quantum CFD (QCFD) [132–135]. In quantum computing, the system can be in a superposition of multiple states at the same time. By adapting conventional algorithms to quantum computing, an enormous increase in speed could be achieved and the use of models for turbulence etc. would become obsolete [136].

Bioreactor [Configuration]	Parameter	Working volume	Software	Reference
Stirred tank bioreactors				
Ambr15 and 250 (Sartorius AG) [2x3BSS, 2xRT6]	k_La	15 mL to 250 mL	Fluent, CFX	[120, 137]
Bioflo IIC (New Brunswick Scientific Co., Inc.) [2xRT6]	k_La	4 L	Fluent	[138]
Biotron GX (Biotron Inc.) [1xRT6, 1xST6, 1xP2]	k_La	2.5 L	Fluent	[139]

Bioreactor [Configuration]	Parameter	Working volume	Software	Reference
CellReady (Merck Millipore) [1xMI]	k_La	1.5 L to 2.5 L	Fluent	[140]
Commercially available single-use bioreactor [1xRT4]	k_La	0.2 m^3 to 1 m^3	CFX	[141]
HyPerforma glass (Thermo Fisher Scientific Inc.) [1xPB4]	k_La	2 L	OpenFOAM	[8]
HyPerforma single-use bioreactor and Dynadrive (Thermo Fisher Scientific Inc.), n. S. [1x3BEE, 3xSD + 1xSD, 2x3BEE]	k_La	2 m^3 to 10 m^3	Fluent	[121]
n. S. [2xRT]	k_La	2 L	Fluent	[142]
n. S. [1xPB6, 1xHPC, 1xRT6, 1xP2, 1xPB2]	k_La	2 L	Fluent	[143]
n. S. [1xRT6, 1xSD, 1xSD]	k_La	10 L	Fluent	[144]
n. S. [3xRT6]	k_La	30 L	CFX	[145]
n. S. [1xMI]	k_La	50 L	Fluent	[146]
n. S. [1xRT6 + 2x3BSS]	k_La	0.2 m^3	Fluent	[147]
n. S. (developed by Guoqiang Bioengineering Equipment Co.) [1xRT4]	k_La	30 mL to 70 mL	CFX	[122]
BIOTECH-5JG (Shanghai Baoxing Bio-Engineer Equipment Co., Ltd.) [1xRT6]	P/V	3 L	Fluent	[148]
BioFlo 415 (Eppendorf SE) [2xRT6]	Θ_m	7 L	CFX	[149]
CellReady (Merck Millipore), n. S. [1xMI]	Θ_m	2 L to 0.2 m^3	Fluent	[150]
n. S. [1xMI]	Θ_m	50 L	Fluent	[151]
n. S. [1xPB4, 1xRT2]	Θ_m	1.5 L	OpenFOAM	[152]
BioFlo 110 (Eppendorf SE) [1xRT6 + 1x3BSS, 2xRT6]	Θ_m, P/V	2.2 L	Autodesk CFD	[153]
n. S. [1xRT6, 1xPB6]	Θ_m, P/V	46.5 L	Simcenter Star-CCM+	[154]
n. S. [1xRT8 + 1RT6]	Θ_m, P/V	54 m^3	M-Star CFD	[155]
n. S. (Sartorius AG) [2x3BSS, 1xRT6 + 1x3BSS]	Θ_m, P/V	15 L	Fluent	[123]
HyPerforma single-use bioreactor and Dynadrive (Thermo Fisher Scientific Inc.) [1xRT6 + 1x3BBS, 3xSD + 1xSD]	Θ_m, P/V, τ	1 L to 50 L	Fluent	[156]
BIOTECH-5JG (Shanghai Baoxing Bio-Engineer Equipment Co., Ltd.) [1xRT6]	k_La, Θ_m	5 L	COMSOL Multiphysics	[157]
n. S. [1x3BSS]	k_La, Θ_m	0.15 m^3 to 0.35 m^3	CFX	[158]
n. S. [4xRT6]	k_La, Θ_m	20 L to 100 m^3	Fluent	[159]
XcellerexTM XDR-10 (Cytiva) [1xPB3]	k_La, Θ_m	4.5 L to 10 L	OpenFOAM, MixIT	[160]
n. S. [4xRT6]	k_La, Θ_m, P/V	38 L	Fluent	[161]
	k_La, Θ_m, P/V	3.9 L to 2 m^3	M-Star CFD	[162]

Bioreactor [Configuration]	Parameter	Working volume	Software	Reference
Xcellerex XDR-200, 500 and 2000 (Cytiva), Applikon 5 L (Applikon Biotechnology B.V.) [1xPB3]				
n. S. [1xRT6+2x3BSS]	k_La, P/V, λ_k	0.3 m^3	Fluent	[124]
n. S. [1xRT6, 2xRT6]	λ_k	3 L and 0.3 m^3	Fluent	[163]
n. S. [1xA310, 1xRT6, 1xPB6]	Sedimentation	24 L	M-Star CFD	[164]
Wave-mixed bioreactors				
Wave bioreactor 20/50 (GE Healthcare)	k_La	5 L	Fluent	[88]
ReadyToProcess WAVE 25 bioreactor (Cytiva)	k_La, P/V, τ	2 L and 20 L	OpenFOAM	[165]
CELL-tainer Utility (Celltainer Biotech BV) [with and w/o expansion channels]	k_La, Θ_m, P/V	0.15 L to 20 L	OpenFOAM	[44]
n. S. (Wave bag)	k_La, Θ_m, P/V	10 L	CFX	[27]
Cellbag CB2L and CB20L (Cytiva)	Sedimentation, τ	2 L and 20 L	Fluent	[166]
Orbitally shaken bioreactors				
TubeSpin 50 (Techno Plastic Products AG)	k_La	10 mL to 50 mL	Fluent, OpenFOAM	[167, 168]
Shake flask (a)	k_La	15 mL to 200 mL	Fluent, CFX, OpenFOAM	[87, 168–170]
TubeSpin 600 (Techno Plastic Products AG)	k_La	100 mL to 500 mL	Fluent	[45]
Orbitally shaken bag (Infors AG)	k_La	1 L to 10 L	OpenFOAM	[168]
n. S. (Cylinder)	k_La	10 mL to 0.1 m^3	Fluent, OpenFOAM	[171, 172]
n. S. (Cylinder)	P/V	10 L to 0.2 m^3	OpenFOAM	[173]
TubeSpin 50 and 600 (Techno Plastic Products AG), Optimum Growth flask 500 and 5000 (Thomson Instrument Company), Shake flask	P/V	10 mL to 3000 mL	OpenFOAM	[10]
SB10-X (Adolf Kühner AG)	k_La, τ	2.5 L to 10 L	Fluent	[174]

The bioreactors were subdivided according to their power input into stirred, wave-mixed and orbitally shaken systems. 3BEE, 3-Blade elephant ear stirrer; 3BSS, 3-Blade segment stirrer; A310, Hydrofoil A310; MI, Marine impeller; P, Paddle stirrer; PB, Pitched blade stirrer; RT, Rushton turbine; SD, Special design; ST, Smith turbine; the number after the label indicates how many stirrer blades were present.

Table 1.
Overview of process characterisations of bioreactors using CFD.

6. Conclusions and outlook

CFD is a widely used tool that has played a role in biotechnology for many years and is constantly gaining in popularity. It can be used to determine classical parameters such as power input, mixing time and oxygen transfer, as well as to carry out further evaluations with spatial and time resolutions. CFD is also an excellent approach for determining hydrodynamic stress. However, depending on the bioreactor system and parameters, different modelling approaches should be selected to determine these parameters. Investigations have shown that for bioreactor modelling, the Fluent commercial software from Ansys dominates, followed by OpenFOAM, an open source solution. Very few authors compute on GPUs, however they still use CPUs. Even tough when used in conjunction with the latest software solutions, GPUs cost less, use less power and perform faster computations. In spite of the considerable computing power currently available, phenomena such as turbulence and gas bubble breakup in industry-relevant simulations, such as bioreactors, are still approximated using semi-empirical models. In the future, more powerful GPUs or even quantum processors for QCFD will increase computational power, and methods like PINN will speed up computations, potentially making it possible to eliminate the need for these models and perform even more accurate and faster simulations. In the second part of the series, the methods described here for process engineering characterisation are demonstrated based on real-world examples, using both commercial and open-source software.

Acknowledgements

We would like to thank Darren Mace for English proofreading.

Conflict of interest

The authors declare no conflict of interest.

Author details

Stefan Seidel*, Cedric Schirmer, Rüdiger W. Maschke, Lia Rossi, Regine Eibl and Dieter Eibl
School of Life Sciences and Facility Management, Institute of Chemistry and Biotechnology, ZHAW Zurich University of Applied Sciences, Wädenswil, Switzerland

*Address all correspondence to: stefan.seidel@zhaw.ch

References

[1] Maschke RW, Pretzner B, John GT, Herwig C, Eibl D. Improved time resolved KPI and strain characterization of multiple hosts in shake flasks using advanced online analytics and data science. Bioengineering. 2022;**9**(8):339. DOI: 10.3390/bioengineering9080339

[2] Seidel S, Maschke RW, Werner S, Jossen V, Eibl D. Oxygen mass transfer in biopharmaceutical processes: Numerical and experimental approaches. Chemie-Ingenieur-Technik. 2021;**93**(1-2):42-61. DOI: 10.1002/cite.202000179

[3] Zeng AP, Deckwer WD. Bioreaction techniques under microaerobic conditions: From molecular level to pilot plant reactors. Chemical Engineering Science. 1996;**51**(10):2305-2314. DOI: 10.1016/0009-2509(96)00087-5

[4] Baeshen MN, Al-Hejin AM, Bora RS, Ahmed MMM, Ramadan HAI, Saini KS, et al. Production of biopharmaceuticals in *E. coli*: Current scenario and future perspectives. Journal of Microbiology and Biotechnology. 2015;**25**(7):953-962. DOI: 10.4014/jmb.1412.12079

[5] BioPlan Associates. Seventeenth Annual Report and Survey of Biopharmaceutical Manufacturing Capacity and Production. Rockville, MD: Technical Report; 2020

[6] Pörtner R. Characteristics of mammalian cells and requirements for cultivation. In: Eibl R, Eibl D, Pörtner R, Catapano G, Czermak P, editors. Cell and Tissue Reaction Engineering. 1st ed. Berlin, Heidelberg: Springer; 2009. pp. 13-53. DOI: 10.1007/978-3-540-68182-3_2

[7] Hossler P, Khattak SF, Li ZJ. Optimal and consistent protein glycosylation in mammalian cell culture. Glycobiology. 2009;**19**(9):936-949. DOI: 10.1093/glycob/cwp079

[8] Seidel S, Eibl D. Influence of interfacial force models and population balance models on the k_La value in stirred bioreactors. Processes. 2021;**9**(7):1185. DOI: 10.3390/pr9071185

[9] Nienow AW. The impact of fluid dynamic stress in stirred bioreactors – the scale of the biological entity: A personal view. Chemie Ingenieur Technik. 2021;**93**(1-2):17-30. DOI: 10.1002/cite.202000176

[10] Maschke RW, Seidel S, Bley T, Eibl R, Eibl D. Determination of culture design spaces in shaken disposable cultivation systems for CHO suspension cell cultures. Biochemical Engineering Journal. 2022;**177**:108224. DOI: 10.1016/j.bej.2021.108224

[11] Jossen V, Eibl R, Eibl D. Single-use bioreactors – an overview. In: Eibl R, Eibl D, editors Single-Use Technology in Biopharmaceutical Manu-facture. Hoboken, NJ: Wiley; 2019. pp. 37-52. DOI: 10.1002/9781119477891.ch4

[12] Schirmer C, Maschke RW, Pörtner R, Eibl D. An overview of drive systems and sealing types in stirred bioreactors used in biotechnological processes. Applied Microbiology and Biotechnology. 2021;**105**(6):2225-2242. DOI: 10.1007/s00253-021-11180-7

[13] Eibl D, Eibl R. Bioreactors for mammalian cells: General overview. In: Eibl R, Eibl D, Pörtner R, Catapano G, Czermak P, editors. Cell and Tissue Reaction Engineering. Berlin, Heidelberg: Springer; 2009. pp. 55-82. DOI: 10.1007/978-3-540-68182-3_3

[14] Werner S, Maschke RW, Eibl D. Eibl R. Bioreactor technology for sustainable

production of plant cell-derived products. In: Pavlov A, Bley T, editors. Bioprocessing of Plant In Vitro Systems. Reference Series in Phytochemistry. Cham: Springer; 2018. DOI: 10.1007/978-3-319-54600-1_6

[15] Schirmer C, Eibl R, Maschke RW, Mozaffari F, Junne S, Daumke R, et al. Single-use technology for the production of cellular agricultural products: Where are we today? Chemie Ingenieur Technik. 2022;**94**: 2018-2025. DOI: 10.1002/cite.202200092

[16] Eibl R, Steiger N, Fritz C, Eisenkrätzer D, Bär J, Müller D, et al. Standardisierter Zellkulturtest zur Identifizierung kritischer Filme. 1st ed. Frankfurt am Main: DECHEMA Gesellschaft für Chemische Technik und Biotechnologie e.V; 2014

[17] Schirmer C, Blaschczok K, Husemann U, Leupold M, Zahnow C, Rupprecht J, et al. Standardized qualification of stirred bioreactors for microbial biopharmaceutical production processes. Chemie Ingenieur Technik. 2017;**89**(12):1766-1772. DOI: 10.1002/cite.201700039

[18] Schirmer C, Dreher T, Leupold M, Glaser R, Castan A, Brown J, et al. Recommendation for Biological Evaluation of Bioreactor Performance for Microbial Processes. 2nd ed. Frankfurt am Main, DE: DECHEMA Gesellschaft für Chemische Technik und Biotechnologie e.V; 2019

[19] Bauer I, Dreher T, Eibl D, Glöckler R, Husemann U, John GT, et al. Recommendations for Process Engineering Characterisation of Single-Use Bioreactors and Mixing Systems by Using Experimental Methods. 2nd ed. Frankfurt am Main, DE: DECHEMA Gesellschaft für Chemische Technik und Biotechnologie e.V; 2020

[20] Jossen V, Eibl D, Eibl R. Numerical methods for the design and description of in vitro expansion processes of human mesenchymal stem cells. In: Herwig C, Pörtner R, Möller J, editors. Digital Twins. 1st ed. Cham: Springer; 2020. pp. 185-228. DOI: 10.1007/10_2020_147

[21] Kaiser SC, Löffelholz C, Werner S, Eibl D. CFD for characterizing standard and single-use stirred cell culture bioreactors. In: Minin IV, Minin OV, editors. Computational Fluid Dynamics Technologies and Applications. London, UK, Rijeka: IntechOpen; 2011. pp. 97-122. DOI: 10.5772/23496

[22] Sagmeister P, Jazini M, Klein J, Herwig C. Bacterial suspension cultures. In: Meyer HP, Schmidhalter DR, editors. Industrial Scale Suspension Culture of Living Cells. Weinheim, Germany: Wiley-VCH Verlag GmbH & Co. KGaA; 2014. pp. 40-93. DOI: 10.1002/9783527683321.ch01

[23] Pörtner R. Bioreactors for mammalian cells. In: Al-Rubeai M editor. Animal Cell Culture. Cham, CH: Springer; 2015. pp. 89-135. DOI: 10.1007/978-3-319-10320-4_4

[24] Nienow AW. Reactor engineering in large scale animal cell culture. Cytotechnology. 2006;**50**(1-3):9-33. DOI: 10.1007/s10616-006-9005-8

[25] Büchs J, Zoels B. Evaluation of maximum to specific power consumption ratio in shaking bioreactors. Journal of Chemical Engineering of Japan. 2001;**34**(5): 647-653. DOI: 10.1252/jcej.34.647

[26] Liu Y, Wang ZJ, Xia JY, Haringa C, Liu YP, Chu J, et al. Application of Euler-

Lagrange CFD for quantitative evaluating the effect of shear force on *Carthamus tinctorius* L. cell in a stirred tank bioreactor. Biochemical Engineering Journal. 2016;**114**:209-217. DOI: 10.1016/j.bej.2016.07.006

[27] Svay K, Urrea C, Shamlou PA, Zhang H. Computational fluid dynamics analysis of mixing and gas–liquid mass transfer in wave bag bioreactor. Biotechnology Progress. 2020;**36**(6):1-10. DOI: 10.1002/btpr.3049

[28] Nienow AW. Impeller selection for animal cell culture. In: Flickinger M, Encyclopedia of Industrial Biotechnology. Hoboken, NJ: John Wiley & Sons, Inc.; 2010. pp. 1-25. DOI: 10.1002/9780470054581.eib636

[29] Sieck JB, Cordes T, Budach WE, Rhiel MH, Suemeghy Z, Leist C, et al. Development of a scale-down model of hydrodynamic stress to study the performance of an industrial CHO cell line under simulated production scale bioreactor conditions. Journal of Biotechnology. 2013;**164**(1):41-49. DOI: 10.1016/j.jbiotec.2012.11.012

[30] Papoutsakis E. Media additives for protecting freely suspended animal cells against agitation and aeration damage. Trends in Biotechnology. 1991;**9**(1): 316-324. DOI: 10.1016/0167-7799(91) 90102-N

[31] Chisti Y. Animal-cell damage in sparged bioreactors. Trends in Biotechnology. 2000;**18**(10): 420-432. DOI: 10.1016/S0167-7799(00) 01474-8

[32] Anane E, Knudsen IM, Wilson GC. Scale-down cultivation in mammalian cell bioreactors - The effect of bioreactor mixing time on the response of CHO cells to dissolved oxygen gradients. Biochemical Engineering Journal. 2021; **166**:107870. DOI: 10.1016/j.bej.2020.107870

[33] Löffelholz C, Kaiser SC, Kraume M, Eibl R, Eibl D. Dynamic single-use bioreactors used in modern liter- and m^3-scale biotechnological processes: Engineering characteristics and scaling up. In: Eibl D, Eibl R, editors. Disposable Bioreactors II. Berlin, Heidelberg: Springer; 2013. pp. 1-44. DOI: 10.1007/10_2013_187

[34] Jossen V, Kaiser SC, Schirmaier C, Herrmann J, Tappe A, Eibl D, et al. Modification and qualification of a stirred single-use bioreactor for the improved expansion of human mesenchymal stem cells at benchtop scale. Pharmaceutical Bioprocessing. 2014;**2**(4):311-322. DOI: 10.4155/pbp.14.29

[35] Rafiq QA, Brosnan KM, Coopman K, Nienow AW, Hewitt CJ. Culture of human mesenchymal stem cells on microcarriers in a 5 l stirred-tank bioreactor. Biotechnology Letters. 2013; **35**(8):1233-1245. DOI: 10.1007/s10529-013-1211-9

[36] Hanga MP, Raga FA, Moutsatsou P, Hewitt CJ, Nienow AW, Wall I. Scale-up of an intensified bioprocess for the expansion of bovine adipose-derived stem cells (bASCs) in stirred tank bioreactors. Biotechnology and Bioengineering. 2021;**118**(8):3175-3186. DOI: 10.1002/bit.27842

[37] Seidel S, Eibl-Schindler R, Eibl D. Laboratory-independent exploration of stirred bioreactors and their fluid dynamics. Education for Chemical Engineers. 2022;**42**:80-87. DOI: 10.1016/j.ece.2022.10.001

[38] Mohamad AA. Lattice Boltzmann Method. London: Springer; 2019. DOI: 10.1007/978-1-4471-7423-3

[39] Wilcox DC. Turbulence Modeling for CFD. 3rd ed. La Cañada, California, USA: DCW Industries, Inc.; 2006

[40] Pathak M, Tamphasna Devi T. CFD investigation of impact of vessel configuration and different impeller types in stirred tank. IOP Conference Series: Materials Science and Engineering. 2020;**1004**(1): 012006. DOI: 10.1088/1757-899X/1004/1/012006

[41] Launder B, Spalding D. The numerical computation of turbulent flows. Computer Methods in Applied Mechanics and Engineering. 1974;**3**(2): 269-289. DOI: 10.1016/0045-7825(74)90029-2

[42] Paschedag AR. Computational fluid dynamics. In: Ley C, Ullmann's Encyclopedia of Industrial Chemistry. Weinheim, Germany: Wiley-VCH Verlag GmbH & Co. KGaA; 2005. pp. 701-720. DOI: 10.1002/14356007.i07_i01

[43] Rodriguez S. Applied Computational Fluid Dynamics and Turbulence Modeling. Cham: Springer International Publishing; 2019. DOI: 10.1007/978-3-030-28691-0

[44] Seidel S, Maschke RW, Kraume M, Eibl-Schindler R, Eibl D. CFD Modelling of a wave-mixed bioreactor with complex geometry and two degrees of freedom motion. Frontiers in Chemical Engineering. 2022;**4**:1-19. DOI: 10.3389/fceng.2022.1021416

[45] Zhu L, Monteil DT, Wang Y, Song B, Hacker DL, Wurm MJ, et al. Fluid dynamics of flow fields in a disposable 600-mL orbitally shaken bioreactor. Biochemical Engineering Journal. 2018;**129**:84-95. DOI: 10.1016/j.bej.2017.10.019

[46] Mastrone MN, Hartono EA, Chernoray V, Concli F. Oil distribution and churning losses of gearboxes: Experimental and numerical analysis. Tribology International. 2020;**151**: 106496. DOI: 10.1016/j.triboint.2020.106496

[47] Wollny S. Experimentelle und numerische Untersuchungen zur Partikelbeanspruchung in gerührten (Bio-) Reaktoren. Berlin: Technische Universität; 2010. [PhD thesis]

[48] Ouro P, Fraga B, Viti N, Angeloudis A, Stoesser T, Gualtieri C. Instantaneous transport of a passive scalar in a turbulent separated flow. Environmental Fluid Mechanics. 2018;**18**(2):487-513. DOI: 10.1007/s10652-017-9567-3

[49] Brackbill J, Kothe D, Zemach C. A continuum method for modeling surface tension. Journal of Computational Physics. 1992;**100**(2): 335-354. DOI: 10.1016/0021-9991(92)90240-Y

[50] Higbie R. The rate of absorption of a pure gas into a still liquid during short periods of exposure. Transactions of the American Institute of Chemical Engineers. 1935;**31**:365-370

[51] Pourtousi M, Sahu J, Ganesan P. Effect of interfacial forces and turbulence models on predicting flow pattern inside the bubble column. Chemical Engineering and Processing: Process Intensification. 2014;**75**:38-47. DOI: 10.1016/j.cep.2013.11.001

[52] Chuang TJ, Hibiki T. Interfacial forces used in two-phase flow numerical simulation. International Journal of Heat and Mass Transfer. 2017;**113**:741-754. DOI: 10.1016/j.ijheatmasstransfer.2017.05.062

[53] Fukuma M, Muroyama K, Yasunishi A. Specific gas-liquid interfacial area and liquid-phase mass transfer coefficient in a slurry bubble column. Journal of Chemical Engineering of Japan. 1987;**20**(3): 321-324. DOI: 10.1252/jcej.20.321

[54] Liao Y, Lucas D. A literature review on mechanisms and models for the coalescence process of fluid particles. Chemical Engineering Science. 2010; **65**(10):2851-2864. DOI: 10.1016/j.ces.2010.02.020

[55] Liao Y, Lucas D. A literature review of theoretical models for drop and bubble breakup in turbulent dispersions. Chemical Engineering Science. 2009; **64**(15):3389-3406. DOI: 10.1016/j.ces.2009.04.026

[56] Nguyen TT, Laurent F, Fox RO, Massot M. Solution of population balance equations in applications with fine particles: Mathematical modeling and numerical schemes. Journal of Computational Physics. 2016;**325**: 129-156. DOI: 10.1016/j.jcp.2016.08.017

[57] Li D, Li Z, Gao Z. Quadrature-based moment methods for the population balance equation: An algorithm review. Chinese Journal of Chemical Engineering. 2019;**27**(3):483-500. DOI: 10.1016/j.cjche.2018.11.028

[58] Jossen V, Pörtner R, Kaiser SC, Kraume M, Eibl D, Eibl R. Mass production of mesenchymal stem cells—impact of bioreactor design and flow conditions on proliferation and differentiation. In: Eberli D editor. Cells and Biomaterials in Regenerative Medicine. London, UK, Rijeka: InTech; 2014. pp. 119-174. DOI: 10.5772/59385

[59] Loubière C, Delafosse A, Guedon E, Chevalot I, Toye D, Olmos E. Dimensional analysis and CFD simulations of microcarrier 'just-suspended' state in mesenchymal stromal cells bioreactors. Chemical Engineering Science. 2019;**203**:464-474. DOI: 10.1016/j.ces.2019.04.001

[60] Loubière C, Delafosse A, Guedon E, Toye D, Chevalot I, Olmos E. Optimization of the impeller design for mesenchymal stem cell culture on microcarriers in bioreactors. Chemical Engineering & Technology. 2019;**42**: 1702-1708. DOI: 10.1002/ceat.201900105

[61] Ding J, Gidaspow D. A bubbling fluidization model using kinetic theory of granular flow. AICHE Journal. 1990; **36**(4):523-538. DOI: 10.1002/aic.690360404

[62] Liu G. Application of the two-fluid model with kinetic theory of granular flow in liquid–solid fluidized beds. In: Kyzas G, Mitropoulos A, editors. Granularity in Materials Science. Rijeka: InTech; 2018. pp. 3-23. DOI: 10.5772/intechopen.79696

[63] Jossen V. Bioengineering aspects of microcarrier-based hMSC expansions in different single-use bioreactors. Berlin: Technische Universität; 2020. [PhD thesis]

[64] Syamlal M, Rogers W, O'Brien T. MFIX documentation theory guide. Technology. 1993;**1004**:1-49. DOI: 10.2172/10145548

[65] Vogt N. Numerische Simulation partikelbeladener Gasströmungen mit der Euler-Lagrange-Methode. Dortmund DE: Technische Universität Dortmund; 2009. [PhD thesis]

[66] Huilier DGF. An overview of the Lagrangian dispersion modeling of heavy particles in homogeneous

isotropic turbulence and considerations on related LES simulations. Fluids. 2021;**6**(4):145. DOI: 10.3390/fluids6040145

[67] Weber A, Bart HJ. Flow simulation in a 2D bubble column with the Euler-Lagrange and Euler-Euler method. The Open Chemical Engineering Journal. 2018;**12**(1):1-13. DOI: 10.2174/1874123101812010001

[68] Zieringer J, Takors R. In silico prediction of large-scale microbial production performance: Constraints for getting proper data-driven models. Computational and Structural Biotechnology Journal. 2018;**16**:246-256. DOI: 10.1016/j.csbj.2018.06.002

[69] Jossen V, Muoio F, Panella S, Harder Y, Tallone T, Eibl R. An approach towards a GMP compliant in-vitro expansion of human adipose stem cells for autologous therapies. Bioengineering. 2020;7(3):77. DOI: 10.3390/bioengineering7030077

[70] Cytiva. Cytodex surface microcarriers. 2020:4. Available from: https://cdn.cytivalifesciences.com/api/public/content/digi-12160-original. [Accessed: November 19, 2022]

[71] Roache PJ. Perspective: A method for uniform reporting of grid refinement studies. Journal of Fluids Engineering. 1994;**116**(3):405-413. DOI: 10.1115/1.2910291

[72] Jasak H. Error analysis and estimation for the finite volume method with applications to fluid flows. Imperial College of Science, Technology and Medicine; 1996. [PhD thesis]

[73] Jiang L, Diao M, Sun H, Ren Y. Numerical modeling of flow over a rectangular broad-crested weir with a sloped upstream face. Water. 2018; **10**(11):1663. DOI: 10.3390/w10111663

[74] Ramírez LA, Pérez EL, García Díaz C, Camacho Luengas DA, Ratkovich N, Reyes LH. CFD and experimental characterization of a bioreactor: Analysis via power curve, flow patterns and $k_L a$. Processes. 2020; **8**(7):878. DOI: 10.3390/pr8070878

[75] Almohammadi KM, Ingham DB, Ma L, Pourkashan M. Computational fluid dynamics (CFD) mesh independency techniques for a straight blade vertical axis wind turbine. Energy. 2013;**58**:483-493. DOI: 10.1016/j.energy.2013.06.012

[76] Werner S, Kaiser SC, Kraume M, Eibl D. Computational fluid dynamics as a modern tool for engineering characterization of bioreactors. Pharmaceutical Bioprocessing. 2014; **2**(1):85-99. DOI: 10.4155/pbp.13.60

[77] Courant R, Friedrichs K, Lewy H. Über die partiellen Differenzengleichungen der mathematischen Physik. In: Math. Ann. 1928;**100**:32–74. DOI: 10.1007/BF01448839

[78] Holzmann T. Mathematics, Numerics, Derivations and OpenFOAM. 7th ed. Bad Wörishofen, DE: Holzmann CFD; 2019. DOI: 10.13140/RG.2.2.27193.36960

[79] Guide for the Verification and Validation of Computational Fluid Dynamics Simulations (G-077-1998). Reston, VA, USA: American Institute of Aeronautics and Astronautics; 2002. DOI: 10.2514/4.472855.001

[80] Xie F, Liu J, Wang J, Chen W. Computational fluid dynamics simulation and particle image velocimetry experimentation of hydrodynamic

performance of flat-sheet membrane bioreactor equipped with micro-channel turbulence promoters with micro-pores. Korean Journal of Chemical Engineering. 2016;**33**(7):2169-2178. DOI: 10.1007/s11814-016-0076-8

[81] de Lamotte A, Delafosse A, Calvo S, Delvigne F, Toye D. Investigating the effects of hydrodynamics and mixing on mass transfer through the free-surface in stirred tank bioreactors. Chemical Engineering Science. 2017;**172**:125-142. DOI: 10.1016/j.ces.2017.06.028

[82] Thomas JMD, Chakraborty A, Berson RE, Shakeri M, Sharp MK. Validation of a CFD model of an orbiting culture dish with PIV and analytical solutions. AIChE Journal. 2017;**63**(9):4233-4242. DOI: 10.1002/aic.15762

[83] Morsi YS, Yang WW, Owida A, Wong CS. Development of a novel pulsatile bioreactor for tissue culture. Journal of Artificial Organs. 2007;**10**(2):109-114. DOI: 10.1007/s10047-006-0369-5

[84] Curran SJ, Black RA. Quantitative experimental study of shear stresses and mixing in progressive flow regimes within annular-flow bioreactors. Chemical Engineering Science. 2004;**59**(24):5859-5868. DOI: 10.1016/j.ces.2004.07.044

[85] Stefan A, Schultz HJ. Use of OpenFOAM® for the investigation of mixing time in agitated vessels with immersed helical coils. In: Nóbrega M, Jasak H, editors. OpenFOAM®. Cham: Springer International Publishing; 2019. pp. 509-520. DOI: 10.1007/978-3-319-60846-4_36

[86] Azizan A, Büchs J. Three-dimensional (3D) evaluation of liquid distribution in shake flask using an optical fluorescence technique. Journal of Biological Engineering. 2017;**11**(1):28. DOI: 10.1186/s13036-017-0070-7

[87] Tianzhong L, Ge S, Jing L, Xiangming Q, Xiaobei Z. Numerical simulation of flow in Erlenmeyer shaken flask. In: Oh HW editor. Computational Fluid Dynamics. Rijeka: InTech; 2010. pp. 157-172. DOI: 10.5772/7102

[88] Zhan C, Hagrot E, Brandt L, Chotteau V. Study of hydrodynamics in wave bioreactors by computational fluid dynamics reveals a resonance phenomenon. Chemical Engineering Science. 2019;**193**:53-65. DOI: 10.1016/j.ces.2018.08.017

[89] Kaiser SC, Werner S, Jossen V, Blaschczok K, Eibl D. Power input measurements in stirred bioreactors at laboratory scale. Journal of Visualized Experiments. 2018;**135**:1-11. DOI: 10.3791/56078

[90] Büchs J, Maier U, Milbradt C, Zoels B. Power consumption in shaking flasks on rotary shaking machines: I. Power consumption measurement in unbaffled flasks at low liquid viscosity. Biotechnology and Bioengineering. 2000;**68**(6):589-593. DOI: 10.1002/(SICI)1097-0290(20000620)68:6⟨589::AID-BIT1⟩3.0.CO;2-J

[91] Shiue A, Chen SC, Jeng JC, Zhu L, Leggett G. Mixing performance analysis of orbitally shaken bioreactors. Applied Sciences. 2020;**10**(16):5597. DOI: 10.3390/app10165597

[92] Bai Y, Moo-Young M, Anderson WA. Characterization of power input and its impact on mass transfer in a rocking disposable bioreactor. Chemical Engineering

Science. 2019;**209**:115183. DOI: 10.1016/j.ces.2019.115183

[93] Panunzi A, Moroni M, Mazzelli A, Bravi M. Industrial case-study-based computational fluid dynamic (CFD) modeling of stirred and aerated bioreactors. ACS Omega. 2022;**7**(29): 25152-25163. DOI: 10.1021/acsomega.2c01886

[94] Schirmer C, Nussbaumer T, Schöb R, Pörtner R, Eibl R, Eibl D. Development, engineering and biological characterization of stirred tank bioreactors. In: Yeh M, Chen Y, editors. Biopharmaceuticals. Rijeka: InTech; 2018. pp. 87-107. DOI: 10.5772/intechopen.79444

[95] Jones SM, Louw TM, Harrison ST. Energy consumption due to mixing and mass transfer in a wave photobioreactor. Algal Research. 2017;**24**:317-324. DOI: 10.1016/j.algal.2017.03.001

[96] Raval K, Kato Y, Büchs J. Comparison of torque method and temperature method for determination of power consumption in disposable shaken bioreactors. Biochemical Engineering Journal. 2007;**34**(3): 224-227. DOI: 10.1016/j.bej.2006.12.017

[97] Villiger TK, Morbidelli M, Soos M. Experimental determination of maximum effective hydrodynamic stress in multiphase flow using shear sensitive aggregates. AICHE Journal. 2015;**61**(5): 1735-1744. DOI: 10.1002/aic.14753

[98] Šrom O, Trávníková V, Wutz J, Kuschel M, Unsoeld A, Wucherpfennig T, et al. Characterization of hydrodynamic stress in ambr250 bioreactor system and its impact on mammalian cell culture. Biochemical Engineering Journal. 2022; **177**:108240. DOI: 10.1016/j.bej.2021.108240

[99] Vivek V, Eka FN, Chew W. Mixing studies in an unbaffled bioreactor using a computational model corroborated with in-situ Raman and imaging analyses. Chemical Engineering Journal Advances. 2022;**9**:100232. DOI: 10.1016/j.ceja.2021.100232

[100] Garcia-Ochoa F, Gomez E. Bioreactor scale-up and oxygen transfer rate in microbial processes: An overview. Biotechnology Advances. 2009;**27**(2): 153-176. DOI: 10.1016/j.biotechadv.2008.10.006

[101] Löffelholz C. CFD als Instrument zur bioverfahrenstechnischen Charakterisierung von single-use Bioreaktoren und zum Scale-up für Prozesse zur Etablierung und Produktion von Biotherapeutika. Brandenburgische Technischen Universität Cottbus; 2013. [Phd thesis]

[102] Lichti M, Bart HJ. Bubble size distributions with a shadowgraphic optical probe. Flow Measurement and Instrumentation. 2018;**60**:164-170. DOI: 10.1016/j.flowmeasinst.2018.02.020

[103] Barigou M, Greaves M. Bubble-size distributions in a mechanically agitated gas—liquid contactor. Chemical Engineering Science. 1992;**47**(8): 2009-2025. DOI: 10.1016/0009-2509(92)80318-7

[104] Laakkonen M, Moilanen P, Alopaeus V, Aittamaa J. Modelling local bubble size distributions in agitated vessels. Chemical Engineering Science. 2007;**62**(3):721-740. DOI: 10.1016/j.ces.2006.10.006

[105] Panckow RP, Comandè G, Maaß S, Kraume M. Determination of particle size distributions in multiphase systems containing nonspherical fluid particles. Chemical Engineering & Technology.

2015;**38**(11):2011-2016. DOI: 10.1002/ceat.201500123

[106] Panckow RP, Reinecke L, Cuellar MC, Maaß S. Photo-optical in-situ measurement of drop size distributions: Applications in research and industry. Oil & Gas Science and Technology – Revue d'IFP Energies Nouvelles. 2017;**72**(3):14. DOI: 10.2516/ogst/2017009

[107] Maaß S, Wollny S, Voigt A, Kraume M. Experimental comparison of measurement techniques for drop size distributions in liquid/liquid dispersions. Experiments in Fluids. 2011;**50**(2):259-269. DOI: 10.1007/s00348-010-0918-9

[108] Heath AR, Fawell PD, Bahri PA, Swift JD. Estimating average particle size by focused beam reflectance measurement (FBRM). Particle & Particle Systems Characterization. 2002;**19**(2):84. DOI: 10.1002/1521-4117(200205)19:2⟨84::AID-PPSC84⟩3.0.CO;2-1

[109] McClure DD, Norris H, Kavanagh JM, Fletcher DF, Barton GW. Validation of a computationally efficient computational fluid dynamics (CFD) model for industrial bubble column bioreactors. Industrial & Engineering Chemistry Research. 2014;**53**(37):14526-14543. DOI: 10.1021/ie501105m

[110] Rotondi M, Grace N, Betts J, Bargh N, Costariol E, Zoro B, et al. Design and development of a new ambr250 bioreactor vessel for improved cell and gene therapy applications. Biotechnology Letters. 2021;**43**(5):1103-1116. DOI: 10.1007/s10529-021-03076-3

[111] Jossen V, Eibl R, Kraume M, Eibl D. Growth behavior of human adipose tissue-derived stromal/stem cells at small scale: Numerical and experimental investigations. Bioengineering. 2018;**5**(4):106. DOI: 10.3390/bioengineering5040106

[112] Kaiser S, Jossen V, Schirmaier C, Eibl D, Brill S, van den Bos C, et al. Fluid flow and cell proliferation of mesenchymal adipose-derived stem cells in small-scale, stirred, single-use bioreactors. Chemie Ingenieur Technik. 2013;**85**(1-2):95-102. DOI: 10.1002/cite.201200180

[113] Delafosse A, Loubière C, Calvo S, Toye D, Olmos E. Solid-liquid suspension of microcarriers in stirred tank bioreactor – experimental and numerical analysis. Chemical Engineering Science. 2018;**180**:52-63. DOI: 10.1016/j.ces.2018.01.001

[114] Zhang W, Zhao F, Li Y, Lou X, Dai C, Lv W, et al. Suspension and transformation performance of poly(2-hydroxyethyl methacrylate)-based anion exchange cryogel beads with immobilized *Lactobacillus paracasei* cells as biocatalysts towards biosynthesis of phenyllactic acid in stirred tank bioreactors. Chemical Engineering Research and Design. 2022;**181**:120-131. DOI: 10.1016/j.cherd.2021.12.010

[115] Sadino-Riquelme C, Hayes RE, Jeison D, Donoso-Bravo A. Computational fluid dynamic (CFD) modelling in anaerobic digestion: General application and recent advances. Critical Reviews in Environmental Science and Technology. 2018;**48**(1):39-76. DOI: 10.1080/10643389.2018.1440853

[116] Haddadi B, Jordan C, Harasek M. Cost efficient CFD simulations: Proper selection of domain partitioning strategies. Computer Physics

Communications. 2017;**219**:121-134. DOI: 10.1016/j.cpc.2017.05.014

[117] Wang M, Tang Y, Guo X, Ren X. Performance analysis of the graph-partitioning algorithms used in OpenFOAM. In: 2012 IEEE Fifth International Conference on Advanced Computational Intelligence (ICACI). Nanjing, China: IEEE; 2012. pp. 99-104. DOI: 10.1109/ICACI.2012.6463129

[118] Pellegrini F, Roman J. Scotch: A software package for static mapping by dual recursive bipartitioning of process and architecture graphs. In: Liddell H, Colbrook A, Hertzberger B, Sloot P. editors. High-Performance Computing and Networking. HPCN-Europe 1996. Lecture Notes in Computer Science, vol 1067. Springer, Berlin, Heidelberg. DOI: 10.1007/3-540-61142-8_588

[119] Harasek M, Horvath A, Jordan C, Kuttner C, Maier C, Nagy J, et al. Steady-state RANS simulation of a swirling, non-premixed industrial methane-air burner using edcSimpleFoam. Project Report. Project Report. Institute of Chemical Engineering, Technische Universität Wien; 2011

[120] T. Tajsoleiman. Automating experimentation in miniaturized reactors [PhD thesis]. Lyngby, DK: Technical University of Denmark; 2018

[121] Scully J, Considine LB, Smith MT, McAlea E, Jones N, O'Connell E, et al. Beyond heuristics: CFD-based novel multiparameter scale-up for geometrically disparate bioreactors demonstrated at industrial 2kL–10kL scales. Biotechnology and Bioengineering. 2020;**117**(6):1710-1723. DOI: 10.1002/bit.27323

[122] Li C, Tian J, Wang W, Peng H, Zhang M, Hang H, et al. Numerical and experimental assessment of a miniature bioreactor equipped with a mechanical agitator and non-invasive biosensors. Journal of Chemical Technology and Biotechnology. 2019;**94**(8):2671-2683. DOI: 10.1002/jctb.6076

[123] Ebrahimi M, Tamer M, Villegas RM, Chiappetta A, Ein-Mozaffari F. Application of CFD to analyze the hydrodynamic behaviour of a bioreactor with a double impeller. Processes. 2019;7(10):694. DOI: 10.3390/pr7100694

[124] Mishra S, Kumar V, Sarkar J, Rathore AS. Mixing and mass transfer in production scale mammalian cell culture reactor using coupled CFD-species transport-PBM validation. Chemical Engineering Science. 2022;**267**:118323. DOI: 10.1016/j.ces.2022.118323

[125] Ansys. Unleashing the power of multiple GPUs for CFD simulations. 2022;**267**

[126] Siemens. Gpu acceleration for cfd simulation, 2022:1-22. Available from: https://blogs.sw.siemens.com/simcenter/gpu-acceleration-for-cfd-simulation/. [Accessed at 01.11.2022]

[127] Posey S, Pariente F. Opportunities for GPU acceleration of OpenFOAM. In: 7th ESI OpenFOAM Conference. Berlin; 15–17 October 2019

[128] Rojek K, Wyrzykowski R, Gepner P. AI-Accelerated CFD Simulation Based on OpenFOAM and CPU/GPU Computing. In: Paszynski M, Kranzlmüller D, Krzhizhanovskaya VV, Dongarra JJ, Sloot PMA, editors. Computational Science – ICCS 2021. ICCS 2021. Lecture Notes in Computer Science. Vol. 12743. Cham: Springer; 2021. DOI: 10.1007/978-3-030-77964-1_29

[129] Vinuesa R, Brunton SL. Enhancing computational fluid dynamics with machine learning. Nature Computational Science. 2022;**2**(6):358-366. DOI: 10.1038/s43588-022-00264-7

[130] Queiroz L, Santos F, Oliveira J, Souza M. Physics-informed deep learning to predict flow fields in cyclone separators. Digital Chemical Engineering. 2021;**1**:100002. DOI: 10.1016/j.dche.2021.100002

[131] Chen D, Gao X, Xu C, Wang S, Chen S, Fang J, et al. FlowDNN: A physics-informed deep neural network for fast and accurate flow prediction. Frontiers of Information Technology & Electronic Engineering. 2022;**23**(2):207-219. DOI: 10.1631/FITEE.2000435

[132] Jóczik S, Zimborás Z, Majoros T, Kiss A. A cost-efficient approach towards computational fluid dynamics simulations on quantum devices. Applied Sciences. 2022;**12**(6):2873. DOI: 10.3390/app12062873

[133] Oz F, Vuppala RKSS, Kara K, Gaitan F. Solving Burgers' equation with quantum computing. Quantum Information Processing. 2022;**21**(1): 30. DOI: 10.1007/s11128-021-03391-8

[134] Gaitan F. Finding solutions of the Navier-Stokes equations through quantum computing—recent progress, a generalization, and next steps forward. Advanced Quantum Technologies. 2021;**4**(10):2100055. DOI: 10.1002/qute.202100055

[135] Chen ZY, Xue C, Chen SM, Lu BH, Wu YC, Ding JC, et al. Quantum approach to accelerate finite volume method on steady computational fluid dynamics problems. Quantum Information Processing. 2022;**21**(4): 137. DOI: 10.1007/s11128-022-03478-w

[136] Jaksch D, Givi P, Daley AJ, Rung T. Variational quantum algorithms for computational fluid dynamics. AIAA Virtual Collection. 2022:1-22

[137] Li X, Scott K, Kelly WJ, Huang Z. Development of a computational fluid dynamics model for scaling-up ambr bioreactors. Biotechnology and Bioprocess Engineering. 2018;**23**(6): 710-725. DOI: 10.1007/s12257-018-0063-5

[138] Valverde MR, Bettega R, Badino AC. Numerical evaluation of mass transfer coefficient in stirred tank reactors with non-Newtonian fluid. Theoretical Foundations of Chemical Engineering. 2016;**50**(6):945-958. DOI: 10.1134/S0040579516060178

[139] Azargoshasb H, Mousavi SM, Jamialahmadi O, Shojaosadati SA, Mousavi SB. Experiments and a three-phase computational fluid dynamics (CFD) simulation coupled with population balance equations of a stirred tank bioreactor for high cell density cultivation. The Canadian Journal of Chemical Engineering. 2016;**94**(1): 20-32. DOI: 10.1002/cjce.22352

[140] Kaiser SC, Eibl R, Eibl D. Engineering characteristics of a single-use stirred bioreactor at bench-scale: The Mobius CellReady 3L bioreactor as a case study. Engineering in Life Sciences. 2011;**11**(4):359-368. DOI: 10.1002/elsc.201000171

[141] Maltby R, Lewis W, Wright S, Smith A, Chew J. Multiphase CFD modelling of single-use-technology bioreactors for industrial biotechnology applications. International Conference on Fluid Flow, Heat and Mass Transfer. 2016; **122**:1-8. DOI: 10.11159/ffhmt16.122

[142] Kerdouss F, Bannari A, Proulx P. CFD modeling of gas dispersion and

bubble size in a double turbine stirred tank. Chemical Engineering Science. 2006;**61**(10):3313-3322. DOI: 10.1016/j.ces.2005.11.061

[143] Moradkhani H, Izadkhah MS, Anarjan N. Experimental and CFD-PBM study of oxygen mass transfer coefficient in different impeller configurations and operational conditions of a two-phase partitioning bioreactor. Applied Biochemistry and Biotechnology. 2017; **181**(2):710-724. DOI: 10.1007/s12010-016-2243-0

[144] Niño L, Gelves R, Ali H, Solsvik J, Jakobsen H. Applicability of a modified breakage and coalescence model based on the complete turbulence spectrum concept for CFD simulation of gas-liquid mass transfer in a stirred tank reactor. Chemical Engineering Science. 2020;**211**: 1-22. DOI: 10.1016/j.ces.2019.115272

[145] Pan A, Xie M, Li C, Xia J, Chu J, Zhuang Y. CFD simulation of average and local gas–liquid flow properties in stirred tank reactors with multiple rushton impellers. Journal of Chemical Engineering of Japan. 2017;**50**(12): 878-891. DOI: 10.1252/jcej.16we242

[146] Amer M, Feng Y, Ramsey JD. Using CFD simulations and statistical analysis to correlate oxygen mass transfer coefficient to both geometrical parameters and operating conditions in a stirred-tank bioreactor. Biotechnology Progress. 2019;**35**(3):1-14. DOI: 10.1002/btpr.2785

[147] Sarkar J, Shekhawat LK, Loomba V, Rathore AS. CFD of mixing of multi-phase flow in a bioreactor using population balance model. Biotechnology Progress. 2016;**32**(3): 613-628. DOI: 10.1002/btpr.2242

[148] Shu L, Yang M, Zhao H, Li T, Yang L, Zou X, et al. Process optimization in a stirred tank bioreactor based on CFD-Taguchi method: A case study. Journal of Cleaner Production. 2019;**230**:1074-1084. DOI: 10.1016/j.jclepro.2019.05.083

[149] Krychowska A, Kordas M, Konopacki M, Grygorcewicz B, Musik D, Wójcik K, et al. Mathematical modeling of hydrodynamics in bioreactor by means of CFD-based compartment model. Processes. 2020;**8**(10):1301. DOI: 10.3390/pr8101301

[150] Wutz J, Waterkotte B, Heitmann K, Wucherpfennig T. Computational fluid dynamics (CFD) as a tool for industrial UF/DF tank optimization. Biochemical Engineering Journal. 2020;**160**:107617. DOI: 10.1016/j.bej.2020.107617

[151] Madhania S, Fathonah NN, Kusdianto TN, Winardi S. Turbulence modeling in side-entry stirred tank mixing time determination. MATEC Web of Conferences. 2021;**333**:02003. DOI: 10.1051/matecconf/202133302003

[152] Oblak B, Babnik S, Erklavec-Zajec V, Likozar B, Pohar A. Digital twinning process for stirred tank reactors/separation unit operations through tandem experimental/computational fluid dynamics (CFD) simulations. Processes. 2020;**8**(11):1-16. DOI: 10.3390/pr8111511

[153] Verma R, Mehan L, Kumar R, Kumar A, Srivastava A. Computational fluid dynamic analysis of hydrodynamic shear stress generated by different impeller combinations in stirred bioreactor. Biochemical Engineering Journal. 2019;**151**:107312. DOI: 10.1016/j.bej.2019.107312

[154] Wu M, Jurtz N, Walle A, Kraume M. Evaluation and application of efficient CFD-based methods for the

multi-objective optimization of stirred tanks. Chemical Engineering Science. 2022;**263**:118109. DOI: 10.1016/j.ces.2022.118109

[155] Haringa C. An analysis of organism lifelines in an industrial bioreactor using Lattice-Boltzmann CFD. Engineering in Life Sciences. 2022;**23**:1-16. DOI: 10.1002/elsc.202100159

[156] Kaiser SC, Decaria PN, Seidel S, Eibl D. Scaling-up of an insect cell-based virus production process in a novel single-use bioreactor with flexible agitation. Chemie Ingenieur Technik. 2022;**94**:1950-1961. DOI: 10.1002/cite.202200103

[157] Jabbari B, Jalilnejad E, Ghasemzadeh K, Iulianelli A. Modeling and optimization of a membrane gas separation based bioreactor plant for biohydrogen production by CFD–RSM combined method. Journal of Water Process Engineering. 2021;**43**:102288. DOI: 10.1016/j.jwpe.2021.102288

[158] Bach C, Yang J, Larsson H, Stocks SM, Gernaey KV, Albaek MO, et al. Evaluation of mixing and mass transfer in a stirred pilot scale bioreactor utilizing CFD. Chemical Engineering Science. 2017;**171**:19-26. DOI: 10.1016/j.ces.2017.05.001

[159] Cappello V, Plais C, Vial C, Augier F. Scale-up of aerated bioreactors: CFD validation and application to the enzyme production by *Trichoderma reesei*. Chemical Engineering Science. 2021;**229**:116033. DOI: 10.1016/j.ces.2020.116033

[160] Kreitmayer D, Gopireddy S, Matsuura T, Kondo S, Kakihara H, Nonaka K, et al. CFD-based characterization of the single-use bioreactor Xcellerex^{TM} XDR-10 for cell culture process optimization. Proceedings of the 5th World Congress on Momentum, Heat and Mass Transfer. 2020;i:1–8. DOI: 10.11159/icmfht20.185

[161] Maluta F, Paglianti A, Montante G. Two-fluids RANS predictions of gas cavities, power consumption, mixing time and oxygen transfer rate in an aerated fermenter scale-down stirred with multiple impellers. Biochemical Engineering Journal. 2021;**166**:107867. DOI: 10.1016/j.bej.2020.107867

[162] Thomas JA, Liu X, DeVincentis B, Hua H, Yao G, Borys MC, et al. A mechanistic approach for predicting mass transfer in bioreactors. Chemical Engineering Science. 2021;**237**:116538. DOI: 10.1016/j.ces.2021.116538

[163] Gelves R, Niño L. CFD prediction of heterogeneities in the scale up of liquid-liquid dispersions. International Journal of Chemical Engineering and Applications. 2014;**5**(2):79-84. DOI: 10.7763/IJCEA.2014.V5.355

[164] Sirasitthichoke C, Teoman B, Thomas J, Armenante PM. Computational prediction of the just-suspended speed, N, in stirred vessels using the lattice Boltzmann method (LBM) coupled with a novel mathematical approach. Chemical Engineering Science. 2022;**251**:117411. DOI: 10.1016/j.ces.2021.117411

[165] Werner S, Stalder M, Perepelitsa N, Eibl D. Engineering characterization of ReadyToProcess WAVE 25 bioreactor system with 20 L Cellbag culture chamber. Uppsala SE: GE Healthcare Bio-Sciences AB; 2016

[166] Öncül AA, Genzel Y, Reichl U, Thévenin D. Flow characterization in wave bioreactors using computational fluid dynamics. In: Proceedings of the 21st Annual Meeting of the European

Society for Animal Cell Technology (ESACT). Dublin, Ireland. June 7-10, 2009. Netherlands, Dordrecht: Springer; 2012. pp. 455–469. DOI: 10.1007/978-94-007-0884-6_78

[167] Zhu LK, Song BY, Wang ZL, Monteil DT, Shen X, Hacker DL, et al. Studies on fluid dynamics of the flow field and gas transfer in orbitally shaken tubes. Biotechnology Progress. 2017;**33**(1):192-200. DOI: 10.1002/btpr.2375

[168] Werner S, Olownia J, Egger D, Eibl D. An approach for scale-up of geometrically dissimilar orbitally shaken single-use bioreactors. Chemie Ingenieur Technik. 2013;**85**(1-2):118-126. DOI: 10.1002/cite.201200153

[169] Zhang H, Williams-Dalson W, Keshavarz-Moore E, Shamlou PA. Computational-fluid-dynamics (CFD) analysis of mixing and gas–liquid mass transfer in shake flasks. Biotechnology and Applied Biochemistry. 2005;**41**(1):1. DOI: 10.1042/BA20040082

[170] Li C, Xia JY, Chu J, Wang YH, Zhuang YP, Zhang SL. CFD analysis of the turbulent flow in baffled shake flasks. Biochemical Engineering Journal. 2013;**70**:140-150. DOI: 10.1016/j.bej.2012.10.012

[171] Zhang X, Bürki CA, Stettler M, De Sanctis D, Perrone M, Discacciati M, et al. Efficient oxygen transfer by surface aeration in shaken cylindrical containers for mammalian cell cultivation at volumetric scales up to 1000L. Biochemical Engineering Journal. 2009;**45**(1):41-47. DOI: 10.1016/j.bej.2009.02.003

[172] Hansen S, Gumprecht A, Micheel L, Hennemann HG, Enzmann F, Blümke W. Implementation of perforated concentric ring walls considerably improves gas-liquid mass transfer of shaken bioreactors. Frontiers in Bioengineering and Biotechnology. 2022;**10**:1-13. DOI: 10.3389/fbioe.2022.894295

[173] Klöckner W, Lattermann C, Pursche F, Büchs J, Werner S, Eibl D. Time efficient way to calculate oxygen transfer areas and power input in cylindrical disposable shaken bioreactors. Biotechnology Progress. 2014;**30**(6):1441-1456. DOI: 10.1002/btpr.1977

[174] Zhu L, Zhao C, Shiue A, Jeng J, Wurm MJ, Raussin G, et al. Fluid dynamics of a pilot-scale OrbShake bioreactor under different operating conditions. Journal of Chemical Technology & Biotechnology. 2022;**97**(4):1027-1036. DOI: 10.1002/jctb.6995

Chapter 4

Computational Fluid Dynamics for Advanced Characterisation of Bioreactors Used in the Biopharmaceutical Industry – Part II: Case Studies

Stefan Seidel, Cedric Schirmer, Rüdiger W. Maschke, Lia Rossi, Regine Eibl and Dieter Eibl

Abstract

The first part of this series on characterisation of bioreactors in the biopharmaceutical industry using computational fluid dynamics presented a literature review to illustrate how characterisation can be performed and which process engineering parameters can be determined using computational fluid dynamics (CFD). In addition, experimental validation methods were presented, and an overview of typical hardware and software was also provided. In this second part, a selection of the authors' research results will be used to demonstrate how the process characterisation of mechanically driven bioreactors for the biopharmaceutical industry can be determined with CFD and then experimentally validated. Three stirred tank bioreactors with different filling volumes and stirrers were used to demonstrate power input and oxygen transfer in single- and two-phase simulations. For wave-mixed and orbitally shaken systems, the fluid flow was transiently simulated and experimentally validated. In addition, the power input was also determined for both systems.

Keywords: hydrodynamic stress, orbitally shaken bioreactor, particle image velocimetry, power input, process engineering characterisation, stirred tank reactor, wave-mixed bioreactor

1. Introduction

Part one of this series showed that process engineering characterisation is essential for understanding and optimising bioreactors and bioprocesses for the biopharmaceutical industry. Process engineering characterisation usually includes determining the specific power input P/V, hydrodynamic stress τ, mixing time Θ_m, volumetric oxygen mass transfer coefficient k_La and, if applicable, the N_{S1} criterion. Determining these parameters by computational fluid dynamics (CFD) and the experimental investigation of the parameters, which serves as validation, were both described in part one. Furthermore, the literature review of 50 publications showed that Ansys Fluent and OpenFOAM are among the most widely used software solutions for CFD applications in biotechnology. Using selected, practice-relevant examples from the authors' research, the second part of the series now demonstrates how process engineering characterisation can be carried out using CFD and then be validated with experimental investigations. The two most frequently used software solutions were selected for the investigations. In Section 2, the specific power input is determined for two stirred reactors with different working volumes and stirrer configurations. This was then validated experimentally by measuring the electric current and the torque, which were then used to determine the specific power input. The determination of the k_La value by coupling population balance modelling and CFD is also presented for a stirred bioreactor and compared with the gassing-out method. In Section 3, the Flexsafe RM 2 L basic single-use bioreactor bag from Sartorius AG is used to demonstrate how the specific power input and the hydrodynamic stress in a wave-mixed bioreactor system can be determined. Particle image velocimetry (PIV) was used for validation. In Section 4, the influence of the contact angle, which is material-dependent, on the specific power input in shake flasks is investigated. For this example, the sickle height was used for validation purposes and compared with experiments and semiempirical formulas.

2. Stirred bioreactor

In this case study, some of the methodological procedures described in part one of this series are used to numerically simulate three different stirred bioreactors to determine the specific power input and the k_La value. The specific power input was determined for two different stirrers (stirrer diameter of 20 mm and 40 mm) according to Bauer et al. [1] for a bearing-free, magnetically driven 2 L benchtop system described by Schirmer et al. [2], based on Levitronix's levitating impeller technology using the BPS-i30 drive (**Figure 1**) to validate the simulations described below. For this purpose, the torque was calculated using the known motor constant K_t with $1.13\ \mathrm{N\ cm\ A^{-1}}$ using eqs (1) and (2) and the required current for stirring:

$$M = K_t \cdot I \tag{1}$$

$$P/V = \frac{2 \cdot \pi \cdot n \cdot M}{V} \tag{2}$$

To determine the specific power input using a numerical simulation, the power input of the stirrer was determined based on the predicted fluid flow (see **Figure 1**) and the acting torque. For this purpose, Ansys Fluent was used with the realisable k-ε-model [3]. Both the vessel walls and the impeller were treated as nonslip boundaries with standard wall functions and the axial velocity at the fluid surface was set to zero. The stirrer rotation was implemented using the multiple reference frame (MRF) method. Discretisation was performed using the first-order upwind scheme and pressure-velocity coupling using the COUPLED algorithm. The fluid domain was defined by an unstructured mesh of approximately $8 \cdot 10^6$ tetrahedrons. The specific power input for the working volumes of 2 L and 4 L was determined for the bacterial

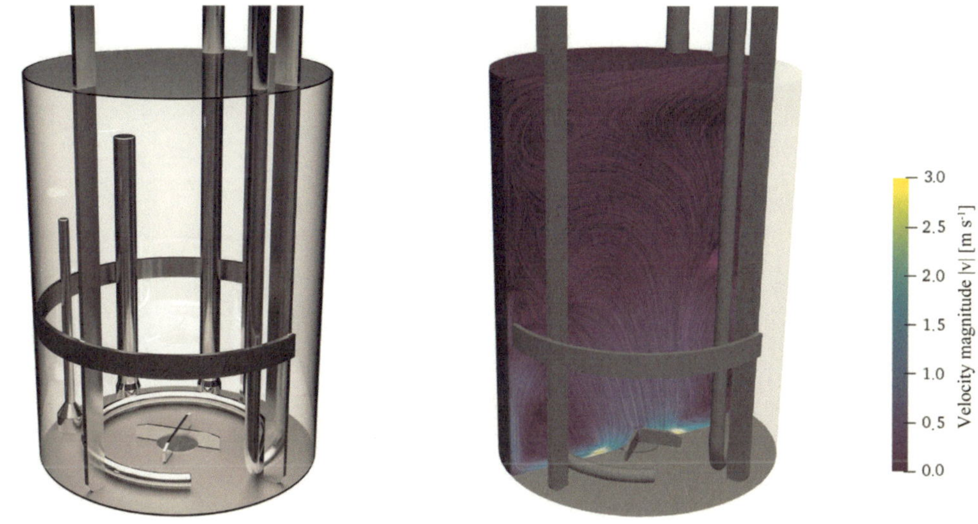

Figure 1.
Computer-aided design (CAD) geometry of the setup and design of the bearing-free, magnetically driven 2 L benchtop system based on Levitronix's levitating impeller technology (left) and numerically derived fluid flow field with a stirrer diameter of 40 mm (right).

version of the Minifors 2 (Infors AG) with a total volume of 6 L, using OpenFOAM's simpleFoam algorithm, the Gauss upwind scheme, the MRF method, and the k-ε-model of Launder and Spalding [4]. Discretisation of the fluid domain was performed with $1.7 \cdot 10^6$ cells (2 L) and $3 \cdot 10^6$ cells (4 L), respectively. Validation of the numerical simulations was performed by determining the specific power input at the same working volumes using a torque meter to which the shaft and the stirrers were attached [1, 5, 6].

The results illustrated in **Figure 2** show good correlation between the experimentally and numerically determined specific power inputs for both the magnetically driven system with the BPS-i30 drive and the Minifors 2. With the exception of the 40-mm diameter stirrer in combination with the BPS-i30, all the other configurations resulted in values >5 kW m^{-3}, making them suitable for microbial applications.

In contrast to the previous ones shown, which were single phase and steady state, the oxygen transfer in another 2 L stirred bioreactor (modified HyPerforma glass bioreactor from Thermo Fisher Scientific Inc.) was determined transiently via a two-phase Euler-Euler simulation [7] (**Figure 3** left). Besides the rotation of the stirrer which was realised via MRF, the aeration was defined by a gas inlet boundary condition at the sparger (a gas outlet is defined at the bioreactor lid). The liquid side mass transfer coefficient k_L value can be calculated according to eq. (3) as a function of the energy dissipation rate (Brüning's [8] adapted formulation was used for these studies). The specific interface a can be calculated according to eq. (4) [9]. Thus, the volumetric oxygen mass transfer coefficient k_La can be determined, by the product of k_L and a. The class method was used to model the gas bubbles and their size distribution with the population balance modelling approach in this example. 24 gas bubble size classes were used, which according to our own investigations and literature values proved to be a good compromise between accuracy and computation time (the computation time increases exponentially as the class number increases) [7, 10, 11]. The model of Laakkonen et al. [12] with binary bubble breakup was used to model bubble

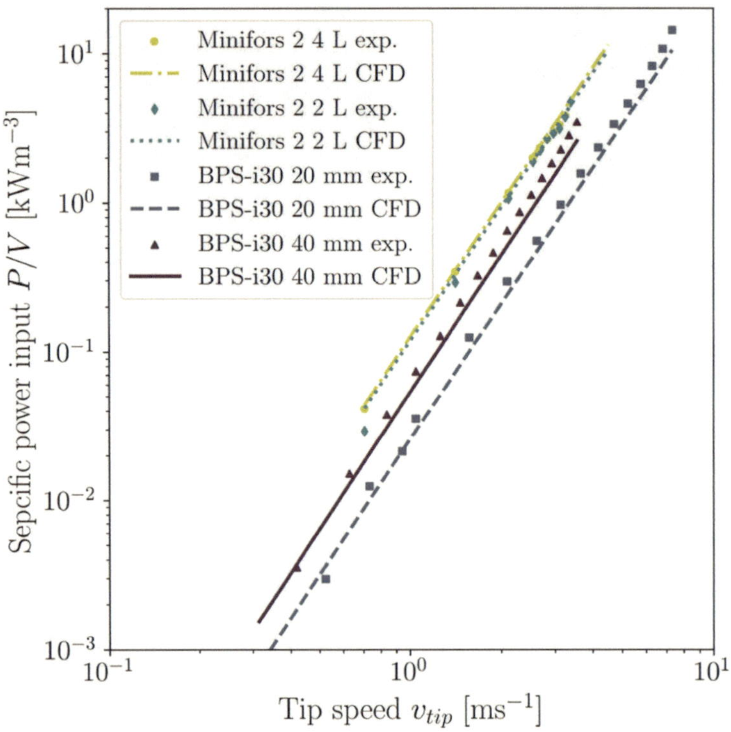

Figure 2.
Double logarithmic representation of the numerically and experimentally determined specific power inputs of the magnetically driven system using the BPS-i30 drive with stirrer diameters of 20 mm and 40 mm as well as of the bacterial version of the Minifors 2 with working volumes of 2 L and 4 L.

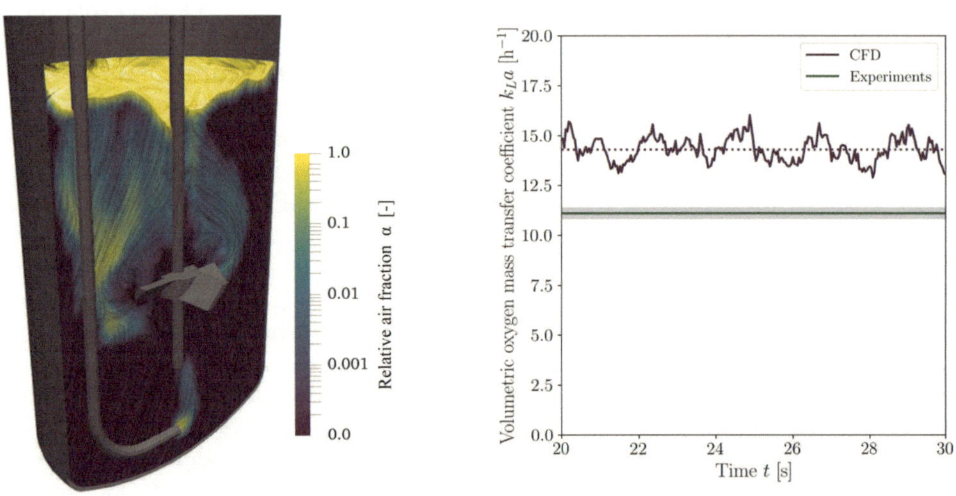

Figure 3.
$k_L a$ value determination by means of CFD coupled with population balance modelling. CFD simulation of the aerated and stirred 2 L HyPerforma glass bioreactor (left). The colour map corresponds to the logarithmised relative gas fraction. The gas flow is visualised using line integral convolution. Temporal variability of the $k_L a$ value determined by CFD simulation and its validation (right). Since the gassing-out method was used for the validation, no temporal statement about the $k_L a$ value is available.

breakup, and the model of Coulaloglou and Tavlarides [13] was used to model the coalescence behaviour. A review of the influence of different coalescence and breakup models can be found in Seidel and Eibl [7]. The interfacial models also have a significant influence on the predicted specific surface area. Different models were investigated by Seidel and Eibl [7], with the model of Schiller and Naumann [14] proving to be the best for calculating the drag force, the model of Tomiyama et al. [15] for calculating the lift force, and the model of Lamb [16] for calculating the virtual mass force. Using this model, a $k_{\mathrm{L}}a$ value of $(14.3 \pm 0.6)\ \mathrm{h}^{-1}$ was determined for a stirrer speed of 600 rpm and an aeration rate of 0.5 vvm (**Figure 3** right). This result was verified experimentally using the gassing-out method according to the DECHEMA e.V. Working Group for Single-Use Technology [1, 17]. An eightfold determination resulted in a measured value of $(11.1 \pm 0.2)\ \mathrm{h}^{-1}$. This deviation can be explained not only by the discretisation error but also much more by the modelling error. The use of various models, some of which are based on empirical approximations, can lead to deviations from the experimental investigations. Only the MRF approach was chosen for the stirrer motion, which is less accurate than the dynamic mesh approach. Nevertheless, this method is suitable for estimating the $k_{\mathrm{L}}a$ value in stirred systems:

$$k_{\mathrm{L}} = \frac{2}{\sqrt{\pi}} \varepsilon \nu \left(\frac{D_{\mathrm{O}_2}}{\nu}\right)^{\frac{1}{4}} \tag{3}$$

$$a = \frac{A_{\mathrm{g}}}{V} = \frac{6\alpha}{d_{32}} \tag{4}$$

3. Wave-mixed bioreactor

The rocker-type motion of a wave-mixed bioreactor can be described by a harmonic oscillation function [18]. This allows the deflection angle φ at time point t to be predicted by eq. (5), whereby φ_{max} corresponds to the maximum deflection angle and $\overline{\omega}$ to the angular velocity, which is calculated using the frequency f (eq. 6). The frequency itself corresponds to the set rocking rate:

$$\varphi_t = \varphi_{max} \cdot \sin(\overline{\omega} t) \tag{5}$$

$$\overline{\omega} = 2\pi \cdot f \tag{6}$$

The motion can be realised with the help of a changing gravity vector $\vec{g}$ ($9.81\ \mathrm{m\ s^{-2}}$) that is decomposed into its x and y components for a given deflection angle (**Figure 4b**). Thus, the gravity vector g_x and g_y can be calculated using eqs (7) and (8). Choosing a maximum deflection angle of 10° and a rocking rate of 25 rpm leads to the outcomes in (**Figure 4c** and **d**):

$$\vec{g}_x = \vec{g} \cdot \sin \varphi_t \tag{7}$$

$$\vec{g}_y = \vec{g} \cdot \cos \varphi_t \tag{8}$$

Another way to mimic the movement of the bag is to rotate or translate the mesh with a user-defined function according to eq. (9):

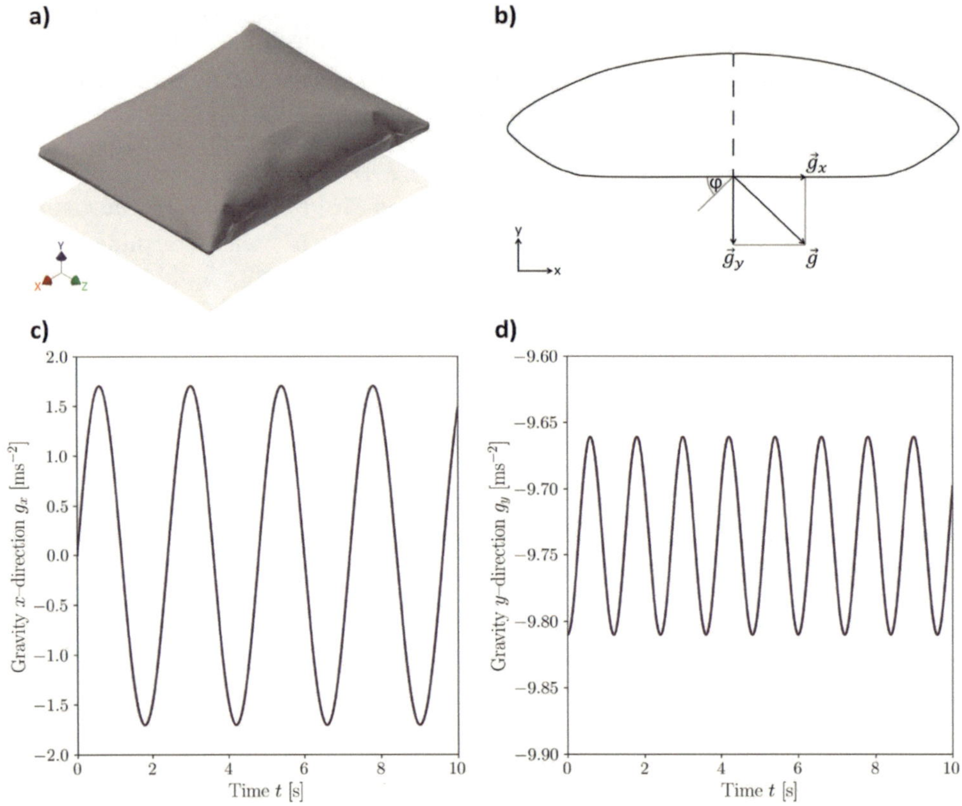

Figure 4.
Representation of a) the 3D-scanned CAD geometry of the Flexsafe RM 2 L basic, b) the decomposition of the gravity vector into its x and y components at a deflection of φ, c) and d) the decomposition of the gravity vector into its x and y components over time at a maximum deflection of 10° and 25 rpm.

$$\overline{\omega}_t = \overline{\omega}\left(\varphi_{max} \cdot \frac{\pi}{180}\right) \cdot \cos(\overline{\omega}t) \quad (9)$$

Both of the above-mentioned methods for implementing the motion in the wave-mixed bioreactor were used for simulations of a two-phase air and water system in a Flexsafe RM 2 L basic single-use bioreactor bag from Sartorius AG (see used CAD geometry in **Figure 4a**). The simulation was performed transiently for liquid volumes of 0.5 L, 1 L, and 1.5 L in Ansys Fluent using velocity-pressure coupling, the SIMPLE algorithm, and the volume of fluid (VOF) model of Hirt and Nichols [19] with two Eulerian phases and a mesh consisting of $1.5 \cdot 10^6$ tetrahedrons. A no-slip assumption was chosen for the walls of the Flexsafe RM. Turbulence modelling was performed using the k-ω model. To achieve sufficient convergence for this procedure, time steps of 10^{-4} s were chosen at the beginning of the simulation.

The CFD simulations were validated by comparing experimentally determined liquid level (**Figures 5** and **6**) and PIV measurements (**Figure 7**) to the simulations. This showed good qualitative and quantitative congruence of the geometric expression of the wave along the bag cross section for different volumes, deflection angles, and rocking rates. The relative deviations remained in the order of magnitude of less than 10% for all of the simulations, which can be attributed to a slight temporal offset of the images being compared. This resulted from the fact that, due to

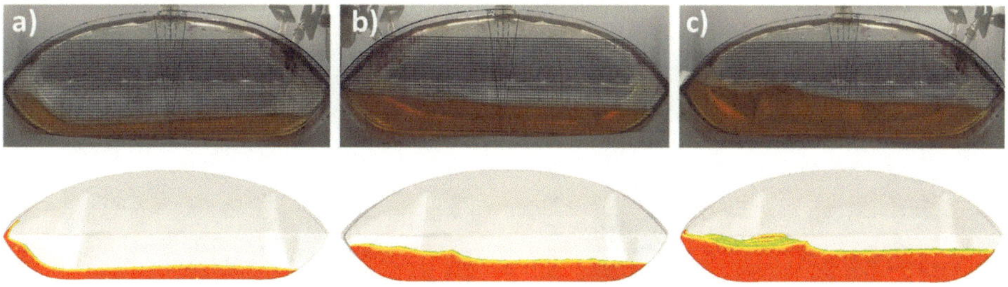

Figure 5.
Qualitative comparison of the fluid surfaces from the fluid level measurements (top) and the numerical simulations (bottom) at a) a fluid volume of 0.5 L, a maximum deflection angle of 6° at 35 rpm, b) a fluid volume of 1 L, a maximum deflection angle of 6° at 25 rpm, and c) a fluid volume of 1.5 L, a maximum deflection angle of 10° at 25 rpm; all images shown have an instantaneous deflection of 3°.

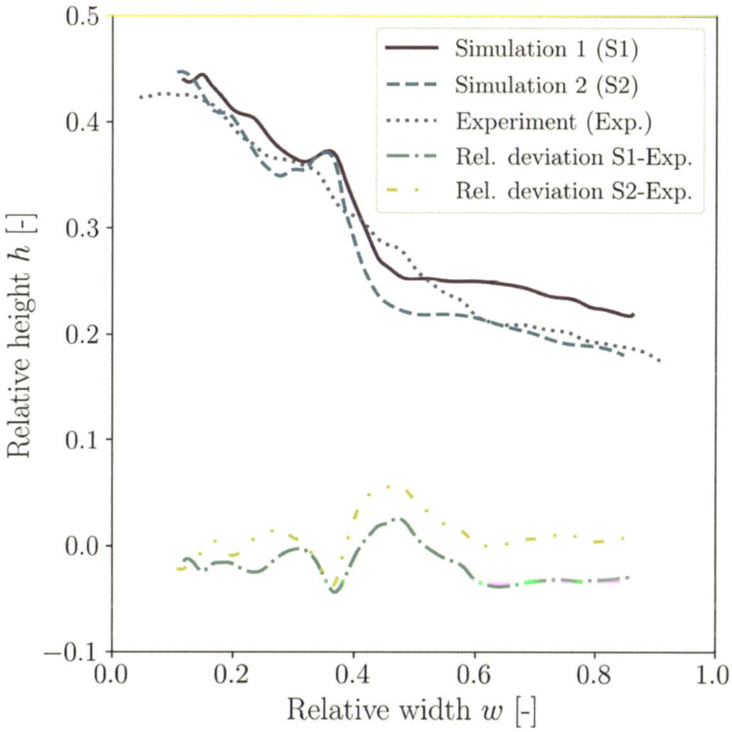

Figure 6.
Dimensionless comparison of the fluid surfaces from the level measurements and the numerical simulations with different motion implementations for a fluid volume of 1 L, a maximum deflection angle of 6° and an instantaneous deflection of 4° at 25 rpm; simulation 1 corresponds to the motion implementation by means of decomposition of the gravity vector into its x and y components; simulation 2 was realised by means of the mesh motion.

the large amounts of data generated during the numerical simulations, the simulation results were only saved every hundredth of a second. As a result, minimal deviations can occur in the comparisons of the wave characteristics at each full degree, because results might not always be available for the exact time point. Another cause of slight deviations is the use of a rigid, 3D-scanned bag shape for the simulations (**Figure 4a**), since in reality slight changes in shape can be observed and, depending on how the bag

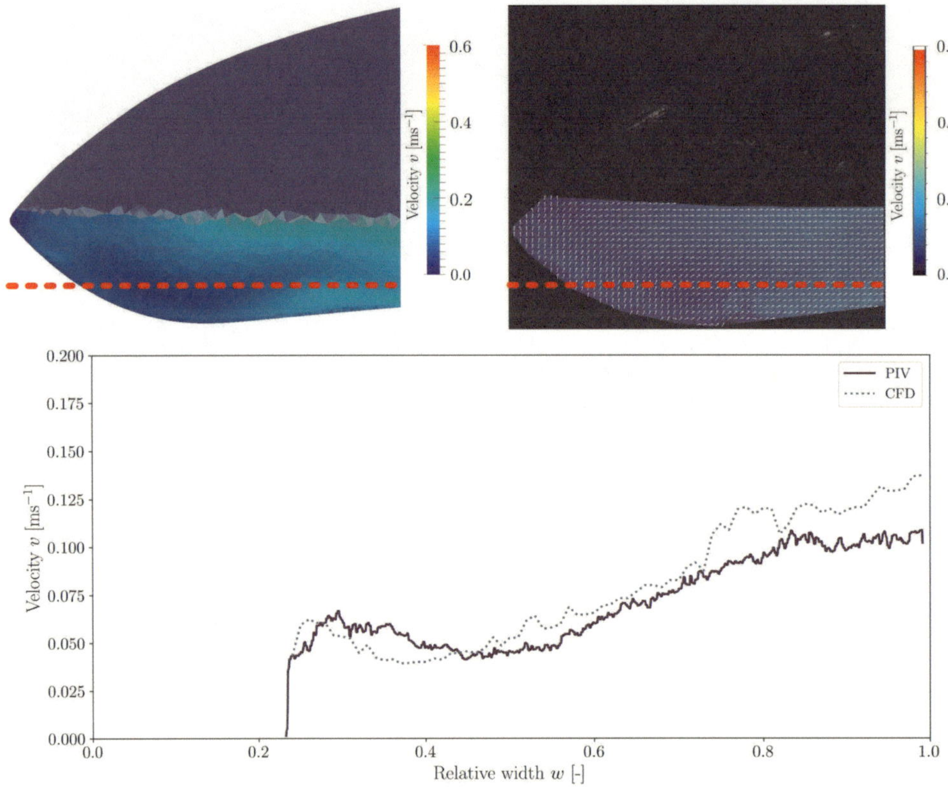

Figure 7.
Qualitative comparison of the fluid velocities from the PIV measurements (top right) and the results of the CFD simulation (top left) for 1 L at 10° and 25 rpm and a quantitative comparison (bottom) of the velocities along the introduced line with a distance of 10 mm above the ground at an instantaneous deflection of 5°; the respective legends indicate the fluid velocities from 0.0 m s^{-1} to 0.6 m s^{-1}.

is fixed, small folds may also form, neither of which were taken into account in the simulation [20].

When comparing the results of the different motion implementations, it becomes clear that only marginal differences occurred between the two simulations (**Figure 6**). The relative deviations of the simulations to each other amounted to a maximum of 5%, which can also be attributed to the previously mentioned reasons as well as to the different mesh handling of Ansys Fluent, since, for example, the change of the mesh position in the Cartesian coordinate system must be taken into account [21]. Nevertheless, this way of handling the mesh appears to result in significantly shorter computation time when opting for the mesh motion option. Computation times of at least 12 weeks were required to reach quasi-stationary periods for gravity vector decomposition, while the computation time for mesh motion was only 4 weeks. Therefore, the latter should be considered the method of choice for future simulations. Validation by means of PIV measurements at specific deflection angles also proved successful. The planes from the PIV measurements were compared to the same planes from the simulations, and the velocity profiles along introduced lines were quantitatively compared. Consequently, good agreement can be assumed to indicate that the simulation results are of high quality. Based on the results, the periodically changing power input and the mean volume-averaged shear stresses could be determined (example shown in

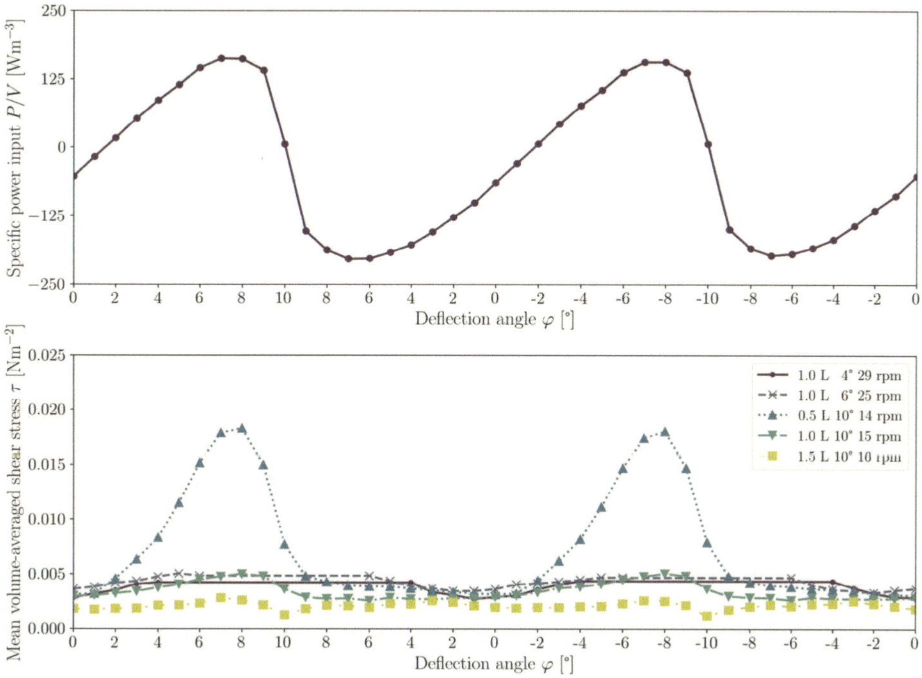

Figure 8.
Characterisation of the specific power input of the numerical simulations at 1 L, 10°, and 25 rpm over an entire period (top) as well as representation of the maximum values of the mean volume-averaged shear stresses over an interval of one period (bottom).

Figure 8), which with values of up to 250 $W\ m^{-3}$ and a maximum of 0.026 $N\ m^{-2}$ indicate good conditions for the cultivation of mammalian cells. For this reason, it comes as no surprise that they are most commonly used for mammalian cell inoculum production processes in feeding and, more recently, perfusion mode [22–25].

4. Orbitally shaken bioreactor

The process engineering and numerical characterisation of orbitally shaken systems poses some challenges for the user. First, the system must be defined. There are different designs of the most common system, the shake flask, be it in shape (Erlenmeyer, Fernbach, Thomson, etc.), sensor installation, diameter of the flask, the neck, or in the material (various plastics and borosilicate glass) [26–29]. In addition to the obvious parameters such as filling height and shaking rate, often neglected aspects such as the shaking amplitude or the contact angle of the material, which depends on the material properties (hydrophilic or hydrophobic), have a decisive influence on the oxygen transfer rate and the power input [30]. This can be explained by the fact that the power is transferred through the vessel wall into the suspension. Likewise, the liquid film that forms on the vessel wall plays a decisive role in the oxygen transfer, as this significantly increases the gas-liquid interface [31]. Finally, in orbitally shaken systems, effects such as the "out of phase" phenomenon can occur, whereas unfavourable cultivation parameters can cause an increase in shaking rate and lead to a decrease in power input, mixing efficiency, and oxygen transfer [32]. Therefore, great care must be taken when choosing simulation parameters and

Figure 9.
Simulation (left) and experiment (right) of the fluid flow in a 500 mL shake flask with 150 mL of working volume at 160 rpm and 25 mm shaking amplitude. The dashed red line indicates the empirically calculated maximum fluid height according to Büchs et al. [33], which deviates by only 2% from the simulated one.

validating simulated orbitally shaken systems. In the first example, the numerically determined fluid motion and the height of the forming sickle are compared with empirical calculations and experiments (**Figure 9**). The simulations were performed using OpenFOAM v9, the VOF method with the interFOAM solver, and the k-ω-SST turbulence model [34]. The CAD model of the 500 mL shake flask was created according to the geometry of Corning Inc.. For glass flasks, ISO 1773 and ISO 4797 could be used [35, 36]. The mesh created by snappyHexMesh consisted of 742,680 mesh cells. The simulation was run transiently for 20 s, using mesh motion (rather than manipulating the force vectors), with a maximum Courant-Friedrichs–Lewy (CFL) number of 0.9. For experimental validation purposes, a flask filled with phenol red coloured water was shaken under identical conditions and the liquid movement was recorded. The calculation was based on an empirical estimation according to Büchs et al. [33], who developed a system of 16 equations to be solved iteratively, that allow the maximal fluid height to be derived. The comparison of the maximum sickle heights showed very good agreement. This was 0.0461 m for the numerical approach, 0.0475 m for the experimental verification, and 0.0469 m for the empirical calculation.

Simulations of sickle height and general fluid motion are ideal for performing an initial verification of the accuracy of a simulation. The second example demonstrates the influence of the contact angle on the power input in shake flasks. The experimental determination of the power input in shake flasks has been described for a wide variety of parameter combinations by Büchs et al. [37, 38]. Glass flasks filled with water were primarily used, with contact angles between 13° and 44° being reported in the literature [39, 40]. **Figure 10** compares experimentally measured data from 250 mL shake flasks (25 mL working volume and 25 mm shaker amplitude) to simulations with contact angles ranging from 30° to 75°. Büchs et al. [37] determined

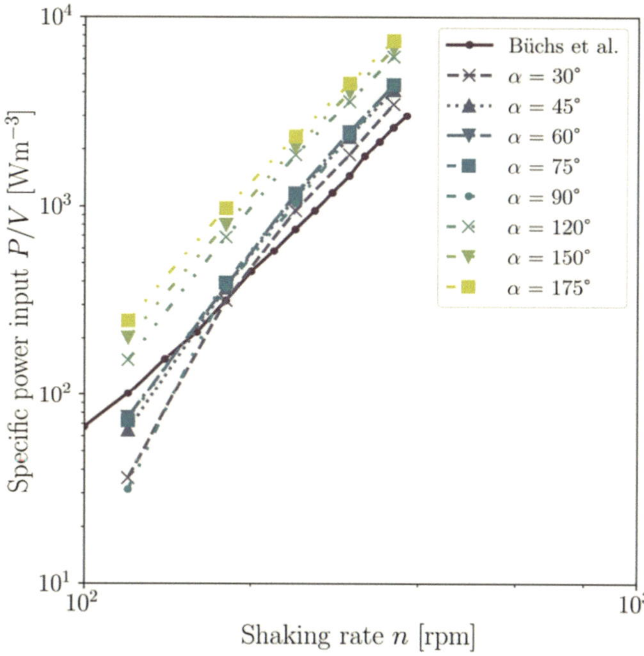

Figure 10.
The specific power input for 250 mL shake flasks, which were shaken with an amplitude of 25 mm and different shaking rates. Different contact angles were investigated by CFD and compared with the experimental data from Büchs et al. [37].

specific power inputs from 0.1 kW m^{-3} to 3 kW m^{-3}, while the results from the CFD simulations ranged from 0.04 kW m^{-3} (30°, 120 rpm) to 4.4 kW m^{-3} (75°, 360 rpm). Good agreement between simulation and measurement was shown at a contact angle of 30°. With higher contact angles, the power input increased, which must be taken into account when changing from glass to (single-use) plastic flasks.

5. Conclusions

The first part of the series illustrated how process engineering parameters for bioreactors used in the biopharmaceutical industry can be determined using CFD and how they can be validated. Furthermore, it was shown that CFD has become well established as a tool for characterising bioreactors, and that both hardware and software are constantly being improved to speed up the calculations and achieve higher accuracy. In this part of the series, practical application examples were used to demonstrate that with the appropriate choice of model, process engineering parameters can be determined to the required degree of accuracy (**Table 1**). This is independent of the mechanical drive for stirred, orbitally shaken, and wave-mixed bioreactor systems. Using different stirred bioreactors, it was possible to show that the calculated specific power input agrees with both the electrical and the torque measurement method. In addition, it was also demonstrated that the k_La value in a stirred system can be calculated in a way that takes coalescence and bubble breakup into consideration, and can be validated with the gassing-out method. Both the free surface and flow velocities in a wave-mixed system were determined using CFD and validated

Bioreactor	Volume	Solver	Turbulence	Mesh in 10^6 cells	Boundary
OpenFOAM					
Minifors 2	2 L[a)] 4 L[a)]	simpleFoam[d)]	*k-ε*	1.7[f)] 3.0[f)]	Wall: no-slip Surface: symmetry
HyPerforma	2 L[b)]	reactingTwo PaseEulerFoam[e)]	*k-ε*	0.3[f)]	Wall: no-slip and Gas in−/outlet
Shake flask	0.025 L[c)] 0.15 L[c)]	interFoam[e)]	*k-ω-SST*	0.09[f)] 0.7[f)]	Wall: contact angle
Fluent					
BPS-i30	2 L[a)]	COUPLED[d)]	*k-ε*	8.0[g)]	Wall: No-slip Surface: symmetry
Flexsafe RM	0.5 L to 1.5 L[c)]	SIMPLE[e)]	*k-ω*	1.5[g)]	Wall: no-slip

a) single-phase b) two-phase Euler-Euler c) VOF d) steady-state e) transient f) mostly hexahedra g) tetrahedrons

Table 1.
Overview of the investigated bioreactor systems with working volumes, solver used, turbulence models, number, and type of mesh cells as well as boundary conditions. The systems are divided into OpenFOAM and fluent cases according to the simulation software used.

experimentally. In addition, the specific power input and the hydrodynamic stress were also determined for this system. Using the example of shake flasks, it was shown that the free surface can also be precisely determined for orbitally shaken systems, with the sickle height of the CFD and experimental validation differing by less than 2%. Furthermore, the influence of the contact angle on the specific power input was also examined. As shown in this series, CFD is used extensively for the characterisation of bioreactors and is likely to be used even more in the future.

Acknowledgements

We would like to thank Sören Werner for contact angle investigations of the shake flasks and Darren Mace for English proofreading.

Conflict of interest

The authors declare no conflict of interest.

Author details

Stefan Seidel*, Cedric Schirmer, Rüdiger W. Maschke, Lia Rossi, Regine Eibl and Dieter Eibl
Institute of Chemistry and Biotechnology, School of Life Sciences and Facility Management, ZHAW Zurich University of Applied Sciences, Wädenswil, Switzerland

*Address all correspondence to: stefan.seidel@zhaw.ch

References

[1] Bauer I, Dreher T, Eibl D, Glöckler R, Husemann U, John GT, Kaiser SC, Kampeis P, Kauling J, Kleebank S, Kraume M, Kuhlmann W, Löffelholz C, Meusel W, Möller J, Pörtner R, Sieblist C, Tscheschke B, and Werner S. Recommendations for Process Engineering Characterisation of Single-Use Bioreactors and Mixing Systems by Using Experimental Methods. Second edition. Tech. rep., Frankfurt am Main. 2020

[2] Schirmer C, Nussbaumer T, Schöb R, Pörtner R, Eibl R, Eibl D. Development, engineering and biological characterization of stirred tank bioreactors. In: Biopharmaceuticals. London, UK, London, UK: InTechOpen; 2018. DOI: 10.5772/intechopen.79444

[3] Shih TH, Liou WW, Shabbir A, Yang Z, Zhu J. A new k-ε eddy viscosity model for high Reynolds number turbulent flows. Computers & Fluids. 1995;**24**(3):227-238. DOI: 10.1016/0045-7930(94)00032-T

[4] Launder B, Spalding D. The numerical computation of turbulent flows. Computer Methods in Applied Mechanics and Engineering. 1974;**3**(2):269-289. DOI: 10.1016/0045-7825(74)90029-2

[5] Kaiser SC, Werner S, Jossen V, Kraume M, Eibl D. Development of a method for reliable power input measurements in conventional and single-use stirred bioreactors at laboratory scale. Engineering in Life Sciences. 2017;**17**(5):500-511. DOI: 10.1002/elsc.201600096

[6] Kaiser SC, Werner S, Jossen V, Blaschczok K, Eibl D. Power input measurements in stirred bioreactors at laboratory scale. Journal of Visualized Experiments. 2018;**135**:1-11. DOI: 10.3791/56078

[7] Seidel S, Eibl D. Influence of interfacial force models and population balance models on the k_La value in stirred bioreactors. PRO. 2021;**9**(7):1185. DOI: 10.3390/pr9071185

[8] >Brüning S. Strömungssimulation als Werkzeug zur Bioreaktorcharakterisierung. PhD thesis. München, DE: Technischen Universität München; 2012

[9] Garcia-Ochoa F, Gomez E. Bioreactor scale-up and oxygen transfer rate in microbial processes: An overview. Biotechnology Advances. 2009;**27**(2):153-176. DOI: 10.1016/j.biotechadv.2008.10.006

[10] Sanyal J, Marchisio DL, Fox RO, Dhanasekharan K. On the comparison between population balance models for CFD simulation of bubble columns. Industrial & Engineering Chemistry Research. 2005;**44**(14):5063-5072. DOI: 10.1021/ie049555j

[11] Venneker BCH, Derksen JJ, Van den Akker HEA. Population balance modeling of aerated stirred vessels based on CFD. AICHE Journal. 2002;**48**(4):673-685. DOI: 10.1002/aic.690480404

[12] Laakkonen M, Alopaeus V, Aittamaa J. Validation of bubble breakage, coalescence and mass transfer models for gas–liquid dispersion in agitated vessel. Chemical Engineering Science. 2006;**61**(1):218-228. DOI: 10.1016/j.ces.2004.11.066

[13] Coulaloglou C, Tavlarides L. Description of interaction processes in agitated liquid-liquid dispersions. Chemical Engineering Science. 1977;**32**(11):1289-1297. DOI: 10.1016/0009-2509(77)85023-9

[14] Schiller L, Naumann A. A drag coefficient correlation. Zeitschrift des Vereins Deutscher Ingenieure. 1935;**77**: 318-320

[15] Tomiyama A, Tamai H, Zun I, Hosokawa S. Transverse migration of single bubbles in simple shear flows. Chemical Engineering Science. 2002; **57**(11):1849-1858. DOI: 10.1016/S0009-2509(02)00085-4

[16] Lamb H. Hydrodynamics. fourth ed. Cambridge: Cambridge University Press; 1993

[17] Seidel S, Maschke RW, Werner S, Jossen V, Eibl D. Oxygen mass transfer in biopharmaceutical processes: Numerical and experimental approaches. Chemie-Ingenieur-Technik. 2021; **93**(1–2):42-61. DOI: 10.1002/cite.202000179

[18] Zhan C, Hagrot E, Brandt L, Chotteau V. Study of hydrodynamics in wave bioreactors by computational fluid dynamics reveals a resonance phenomenon. Chemical Engineering Science. 2019;**193**:53-65. DOI: 10.1016/j.ces.2018.08.017

[19] Hirt CW, Nichols BD. Volume of fluid (VOF) method for the dynamics of free boundaries. Journal of Computational Physics. 1981;**39**(1): 201-225. DOI: 10.1016/0021-9991(81)90145-5

[20] Seidel S, Maschke RW, Kraume M, Eibl-Schindler R, Eibl D. CFD modelling of a wave-mixed bioreactor with complex geometry and two degrees of freedom motion. Frontiers in Chemical Engineering. 2022. 1-19. DOI: 10.3389/fceng.2022.1021416

[21] ANSYS Inc. ANSYS Fluent Theory Guide - Release 2022 R2. Canonsburg: ANSYS Inc; 2022;**4**:1021416

[22] Pörtner R. Bioreactors for Mammalian Cells. In: Al-Rubeai M, editor. Animal Cell Culture. Cell Engineering. Cham: Springer; 2015;**9**:89-135. DOI: 10.1007/978-3-319-10320-4_4

[23] Neunstoecklin B, Stettler M, Solacroup T, Broly H, Morbidelli M, Soos M. Determination of the maximum operating range of hydrodynamic stress in mammalian cell culture. Journal of Biotechnology. 2015;**194**:100-109. DOI: 10.1016/j.jbiotec.2014.12.003

[24] Keane JT, Ryan D, Gray PP. Effect of shear stress on expression of a recombinant protein by Chinese hamster ovary cells. Biotechnology and Bioengineering. 2003;**81**(2):211-220. DOI: 10.1002/bit.10472

[25] Schirmer C, Müller J, Steffen N, Werner S, Eibl R, Eibl D. How to produce mAbs in a cube-shaped stirred single-use bioreactor at 200 L scale. In: Animal Cell Biotechnology. NY, New York: Humana New York; 2020. pp. 169-186. DOI: 10.1007/978-1-0716-0191-4_10

[26] Winkler K, Socher ML. Shake flask technology. In: Encyclopedia of Industrial Biotechnology. Hoboken, NJ, USA: John Wiley & Sons, Inc.; 2014. pp. 1-16. DOI: 10.1002/9780470054581.eib651

[27] Maschke RW, Seidel S, Bley T, Eibl R, Eibl D. Determination of culture design spaces in shaken disposable cultivation systems for CHO suspension cell cultures. Biochemical Engineering Journal. 2022a;**177**:108224. DOI: 10.1016/j.bej.2021.108224

[28] Maschke RW, John GT, Eibl D. Monitoring of oxygen, pH, CO_2, and biomass in smart single-use shake flasks. Chemie Ingenieur Technik. 2022b;**94**:

1995-2001. DOI: 10.1002/cite.2022 00094

[29] Suresh S, Srivastava V, Mishra I. Critical analysis of engineering aspects of shaken flask bioreactors. Critical Reviews in Biotechnology. 2009;**29**(4): 255-278. DOI: 10.3109/ 07388550903062314

[30] Maier U, Büchs J. Characterisation of the gas–liquid mass transfer in shaking bioreactors. Biochemical Engineering Journal. 2001;7(2):99-106. DOI: 10.1016/ S1369-703X(00)00107-8

[31] Kato Y, Tada Y, Iwanaga E, Nagatsu Y, Iwata S, Lee YS, et al. Effects of liquid film formed on flask surface on oxygen transfer rate in shaking flask and development of baffled shaking vessel by optical method based on sulfite oxidation. Journal of Chemical Engineering of Japan. 2005;**38**(11): 873-877. DOI: 10.1252/jcej.38.873

[32] Büchs J, Lotter S, Milbradt C. Out-of-phase operating conditions, a hitherto unknown phenomenon in shaking bioreactors. Biochemical Engineering Journal. 2001;**7**(2):135-141. DOI: 10.1016/S1369-703X(00)00113-3

[33] Büchs J, Maier U, Lotter S, Peter CP. Calculating liquid distribution in shake flasks on rotary shakers at waterlike viscosities. Biochemical Engineering Journal. 2007;**34**(3):200-208. DOI: 10.1016/j.bej.2006.12.005

[34] Menter F. Zonal two equation $k - \omega$ turbulence models for aerodynamic flows. In: 23rd Fluid Dynamics, Plasmadynamics, and Lasers Conference. Reston, Virigina: American Institute of Aeronautics and Astronautics; 1993. DOI: 10.2514/ 6.1993-2906

[35] International Organization for Standardization. DIN ISO 1773:1999-05 Laboratory Glassware - Narrow-Necked Boiling Flasks, 1999

[36] International Organization for Standardization. ISO 4797:2015 Laboratory Glassware — Boiling Flasks with Conical Ground Joints, 2015

[37] Büchs J, Maier U, Milbradt C, Zoels B. Power consumption in shaking flasks on rotary shaking machines: I. power consumption measurement in unbaffled flasks at low liquid viscosity. Biotechnology and Bioengineering. 2000a;**68**(6):589-593. DOI: 10.1002/ (SICI)1097-0290(20000620)68: 6⟨589::AID-BIT1⟩3.0.CO;2-J

[38] Büchs J, Maier U, Milbradt C, Zoels B. Power consumption in shaking flasks on rotary shaking machines: II. Nondimensional description of specific power consumption and flow regimes in unbaffled flasks at elevated liquid viscosity. Biotechnology and Bioengineering. 2000b;**68**(6):594-601. DOI: 10.1002/(SICI)1097-0290 (20000620)68:6⟨594::AID-BIT2⟩3.0. CO;2-U

[39] Vogler EA. Structure and reactivity of water at biomaterial surfaces. Advances in Colloid and Interface Science. 1998;**74**(1–3):69-117. DOI: 10.1016/S0001-8686(97)00040-7

[40] Mittal K, editor. Advances in Contact Angle, Wettability and Adhesion. Hoboken, NJ, USA: John Wiley & Sons, Inc.; 2015. DOI: 10.1002/ 9781119117018

Chapter 5

The Effect of Al_2O_3 Concentration in Annular Fuels for a Typical VVER-1000 Core

Valiyollah Ghazanfari

Abstract

In this study, the effect of Al_2O_3 with various volume percentages in annular fuels for a typical VVER-1000 core is investigated. CFD is used to simulate the fuel rod. The results of the present study are compared with those of other published studies. Furthermore, the mentioned analytical methods are validated by comparing the final safety analysis report (FSAR). By comparing the results, it was found that the results are in good agreement with other studies. The results show that using annular fuel instead of solid fuel in the core of the reactor will increase the security and efficiency of the nuclear power plant. By increasing the volumetric concentration of nanofluid from 0.1 to 0.2, the rate in decrease of the peak temperature in the central fuel increases from 9% to 13%. Moreover, the clad temperature decreases from 626 K to 620 K (the clad temperature with pure water is 635 K).

Keywords: annular fuel, nanofluid, VVER-1000, heat transfer, coolant

1. Introduction

In recent years, nanofluids in various industries have been considered because of their ability to improve heat transfer. Nanoparticles are typically used in the form of nanofluids that consist of nanoparticles and a base fluid. The effective thermal conductivity is increased by the nanoparticles in a nanofluid, which can significantly improve heat transfer [1]. Today, nanoparticles are present in nearly all science and engineering branches. Nanofluids are colloidal dispersions of nanometer-sized materials (stable metals, metal oxides, oxide ceramics, metal carbides, etc.). In particular, Al_2O_3 nanofluids have been widely investigated and used due to their high productivity, higher thermal properties, and stability [2]. Accordingly, many studies have been done on Al_2O_3 nanofluids in heat transfer applications [3, 4].

The characteristics of thermophysical for Al_2O_3 are essential to be examined to find how Al_2O_3 nanoparticles influence the heat transfer and hydraulic properties of nanofluids [5]. Pak and Cho [6] measured the thermophysical characteristics of TiO_2 and Al_2O_3 nanofluids. Viscosity was determined by changing the volume fraction from 1–10%. The results prove that the density and viscosity increase with the volume fraction. Also, viscosity is approximately independent of the shear rate. Agarwal et al.

[7] produced Al_2O_3 nanoparticles and analyzed thermal conductivity. It is dispersed in ethylene glycol or water with different concentrations. At a temperature of 1000 C, the synthesized Al_2O_3 nanoparticles are primarily stable and improve thermal conductivity. Kiruba et al. [8] investigated the effect of adding polyethylenimine on the rheological properties of Al_2O_3 nanofluids. It was proposed that adding polyethylenimine makes the viscosity independent of the fluid temperature and improves gel formation. Kumar et al. [9] investigated the effect of temperature changes (30 C, 40 C, 50 C) on the thermal conductivity of Al_2O_3 nanofluid. The results showed that increasing the volume fraction of nanoparticles in water from 0.01 to 0.08% vol.% improves thermal conductivity, and with increasing temperature, thermal conductivity is further improved.

In a nuclear reactor, heat is released through fission in the fuel rod. Heat is then transferred to the cladding through thermal conduction from the fuel surface [10]. From the cladding surface, heat is transferred to the coolant through convection. The heated coolant is passed to heat exchangers, where steam is generated to operate a power cycle. Nuclear fuel rods are employed as heat sources. Most nuclear reactors run on fuel rods containing the uranium-238 and uranium-235 isotopes. Fuel fission rate and thermal neutron flux affect heat generation in a nuclear reactor. Also, by changing the mass flow rate of the reactor coolant and the temperature difference across the core, the thermal power produced by a reactor changes directly [11]. Uranium is usually in the form of pure metal, in a compound such as $U0_2$, uranium oxide, or in the form of an alloy with another metal such as aluminum or zirconium (in the form of rectangular plates or long cylindrical rods).

Optimal properties of a fuel that must be fissile include high thermal conductivity, good corrosion resistance, good mechanical resistance at high temperatures, good corrosion resistance, and a high limiting temperature for operation [12].

The fuel rod surface heat flux is the most critical factor in developing or designing a new fuel rod because it must be measured as the maximum fuel centerline temperature. As shown in **Figure 1**, the generated heat in the fuel pellet flows into a coolant through only one rod surface in the solid-type fuel, whereas the annular fuel has two surfaces and a dual coolant channel. Thus, the heat generated in the annular fuel can flow into the inner or outer surface.

The internally and externally cooled annular fuel increases power density in the standard Westinghouse PWR while keeping or increasing the safety margin [13, 14]. In 2017, the thermal–hydraulic analyses of externally and internally cooled annular fuel were investigated. This study identified the geometry that allows the largest possible power density while maintaining or increasing the minimum departure from

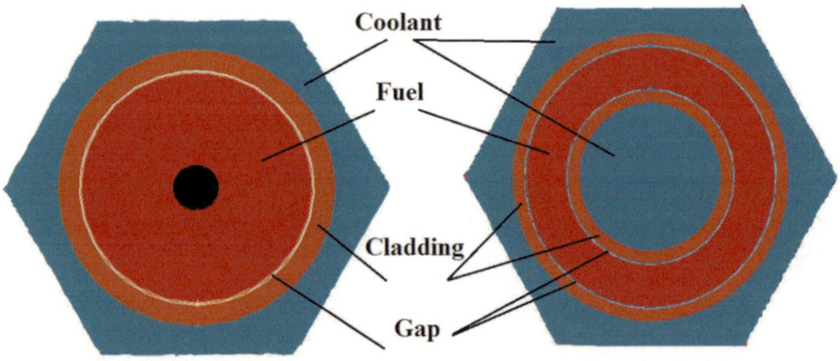

Figure 1.
Schematic of solid (left) and annular fuel (right).

nucleate boiling ratio (MDNBR) margin in current PWRs [15]. The DUO THERM program investigated the design analysis of a dual-cooled annular fuel. The program studied pellet and cladding deformations caused by irradiation and power variation to predict the inner and outer heat fluxes and fuel temperature. Using the DUO THERM program, the fuel temperature and heat flux were estimated for a reference annular fuel design. The results showed that the annular fuel heat flux was significantly influenced by the inner and outer gap conductance behaviors. At zero burnup, the heat flux of the inner cladding was maximum. Furthermore, the location of maximum fuel temperature was altered by gap conductance asymmetry [16].

KAERI (Korea Atomic Energy Research Institute) developed a power-uprated annular fuel reloading. This project aimed to develop annular fuel that can be used in the current OPR-1000 pressurized water reactor. It was considered a 12 × 12 annular fuel to evaluate the thermal–hydraulic performance and compared its efficiency against a 16 × 16 cylindrical fuel assembly. The aim of this study was to evaluate dual-cooled annular fuel at normal power for OPR-1000 and finally to measure the possibility of 120% core power. The results illustrated that annular fuel has sufficient margin available on DNB and fuel pellet temperature relative to cylindrical fuel [17, 18]. Two significant safety coefficients (prompt reactivity and power coefficients) of a typical PWR were considered for the annular fuel core using the MCNP-5 code. The optimized 13 × 13 arrays performed the calculations for annular configuration [19]. Moreover, an internally and externally cooled annular fuel was simulated and adapted to a typical VVER-1000 reactor. The results demonstrated that an annular pin configuration, called annular-8, was suggested based on the fully neutronics and MDNBR evaluations [20].

One of the main components of a nuclear reactor is the cooling system. The coolant enters the core at a low temperature and exits at a high temperature after the fission energy is transferred to it. Heat is then transferred from the high-temperature fluid to other thermodynamic components, and eventually, electrical power is generated. High-performance cooling plays a vital role in the efficiency and security of the nuclear power plant. Consequently, the investigation of the effects of nanofluids has been one of the important subjects in recent years. Nanofluids have been found to possess improved thermo-physical properties such as thermal conductivity. Several studies have shown that nanofluids have great potential for increasing heat transfer rates in various application cases while incurring either little or no penalty in pressure drop [21–23].

In an experimental study, it was investigated different volume concentrations of Al_2O_3 nanofluid flowing in a horizontal shell and tube heat exchanger. The results show that the nanofluid's convective heat transfer coefficient is slightly higher than the base fluid at the same mass flow rate [24]. Also, the nanofluids were simulated as the coolant in the VVER-1000 reactor core to analyze thermal–hydraulic performance using the porous media approach. The results displayed that the temperature of the coolant increases with the concentration of nanoparticles. Due to nanofluids' higher heat transfer coefficient than pure water, the coolant flow rate can be reduced [25].

In the present study, the effect of Al_2O_3 with various volume percentages in annular fuels on a typical VVER-1000 core was investigated. The FLUENT 6.3.26 code is used together with Gambit mesh generation software to model the annular fuels. The prediction–correction method with a SIMPLEC algorithm is applied to numerical solutions. The k-ε model is used to consider the turbulence effect. The cosine form of generated heat in the axial direction of the fuel rod is defined using the User Defined Function (UDF). Firstly, the simulation accuracy is proved by comparing with other studies and the final safety analysis report (FSAR) for the solid fuel with the concentration of nanoparticles. Then, it is simulated annular fuels with various volume percentages of nanofluid. Finally, the temperature distribution in the fuel, clad, and coolant with nanoparticles and pure water concentration is presented.

2. Material and methods

It is essential to control the diameter tolerance of annular fuel for actualizing dual cooled fuel. Therefore, a comprehensive calculation of the optimal dimensions and array of annular fuel rods for a VVER-1000 is applied based on the moderator to fuel ratio, fuel rod pitch, and annular heat surface area to reference. Furthermore, some thermal–hydraulic limitations are considered to achieve a suitable configuration for the annular fuel rods. The annular-8 configuration is nominated as the most promising configuration based on fully thermal–hydraulic analysis and neutronic research [19].

There are 163 fuel assemblies in the core of the VVER-1000 reactor. It has been arranged in a hexagonal lattice with a lattice pitch of 23.6 cm. There are 311 fuel rods, 18 guiding channels for control rods and/or burnable absorber rods (BARs), and a central channel in each fuel assembly. **Figure 2** shows the arrangement of the fuel assembly [26].

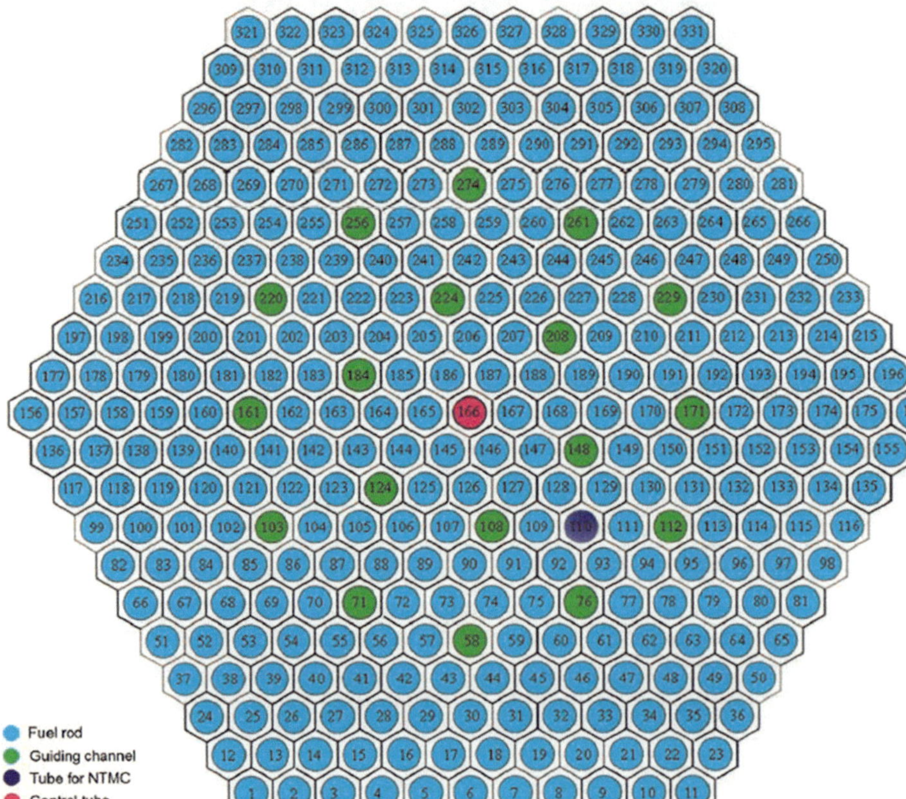

Figure 2.
The arrangement of fuel assembly [14, 15].

Table 1 presents significant specifications for the core of the VVER-1000 reactor.

The typical layout of a fuel assembly for the annular case with 8 × 8 arrays is demonstrated in **Figure 3**. There are 156 fuel rods and 12 guiding channels for control rods in each fuel assembly [20].

Table 2 shows the details of the designed values of the annular fuel which have been calculated by Mozafari (2013) [20]. The MNCP5 and COBRA-EN codes were used to find many neutronics and thermo-hydraulics core parameters.

In the present study, the geometry is drawn with GAMBIT software. GAMBIT offers a concise and powerful set of solid modeling-based geometry tools. If you already employ a CAD package, GAMBIT runs both the geometry import and "clean-up" functions that you'll require. Top down geometry construction using 3D primitives without the complexity of a full-fledged CAD package allows you to create geometries fast. Different CFD problems need different mesh types, and GAMBIT gives you all the options you need in a single package.

In this study, triangular cells are employed to generate the meshes. The number of computational cells is 1.01 million. **Figure 4** displays the quality of the mesh in the fuel rod.

FLUENT reads the generated mesh using GAMBIT. FLUENT is the world's largest commercial Computational Fluid Dynamics (CFD) software. FLUENT is a Green-Gauss Finite Volume Method with a Cell-Centered formulation. The major point is the finite volume method (FVM).

Parameter	Parameter	Value
Core	Reference pressure (MPa)	15.7
	Reactor thermal power (MWt)	3000
	Inlet coolant flow rate (m^3/h)	84,800
	Inlet coolant enthalpy (kJ/kg)	1290
	Coolant temperature at the core inlet (K)	561.15
Fuel assembly	Fuel assembly form	Hexagonal
	Number of fuel assembly in the core	163
	Pitch between the assemblies	23.6
	Number of fuel rod in the fuel assembly	311
	Fuel rod pitch (mm)	12.75
Fuel rod	Hole diameter in the fuel pellet (mm)	1.5
	Fuel pellet outside diameter (mm)	7.57
	Cladding inner diameter (mm)	7.73
	Cladding outer diameter (mm)	9.1
	Fuel pellet material	UO2
	Cladding material	Alloy Zr + 1% Nb
	Fuel pellet density (g/cm^3)	10.4–10.7

Table 1.
VVER-1000 reactor specifications [15].

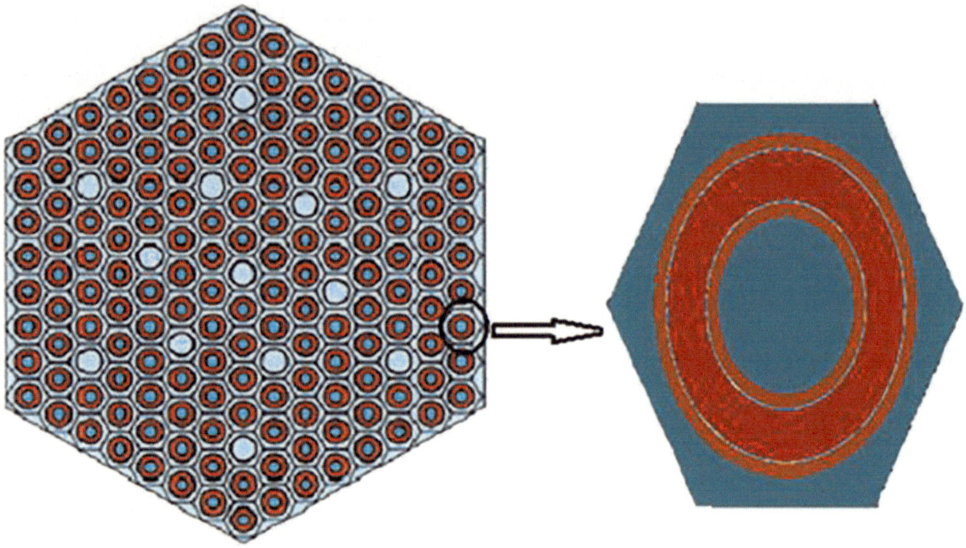

Figure 3.
The fuel assembly layout of annular pins and equivalent lattice cell [8].

Type	Value
Fuel pellet outside radius (mm)	6.920
Outer cladding outer radius (mm)	7.685
Outer cladding thickness (mm)	0.685
Inner cladding inner radius (mm)	3.906
Inner cladding thickness (mm)	0.685
Outer gap width (mm)	0.08
Inner gap width(mm)	0.08
Predicted pitch (mm)	17.77

Table 2.
Dimensions of the Annular-8 fuel rods [8].

The same boundary conditions, such as heat flux, outlet pressure, inlet temperature, and mass flow rate, are considered for nanofluids and pure water. The boundary conditions for the inlet of the fuel assembly, temperature, T_0 = 561 K and profiles of uniform axial velocity,u_0 are set. The non-slip conditions are considered for the fuel rod wall. The rate of 1.753 kg/m.s is considered for the net flow in the fuel assembly. Pressure is set at 15.7 MPa during reactor operation.

$q = q_{max} \cos(\pi z/L)$ is the axial heat-generation distribution, where q, z, and L are the volumetric heat-generation, the axial length, and the active fuel rod length, respectively. To define the cosine function, it is written in the C programming language and compiled in FLUENT.

The governing equations of the nanofluid for the conservation of mass, momentum, and energy in a steady state are written below [27, 28]:

$$\nabla.\left(\rho_{nf} V\right) = 0 \tag{1}$$

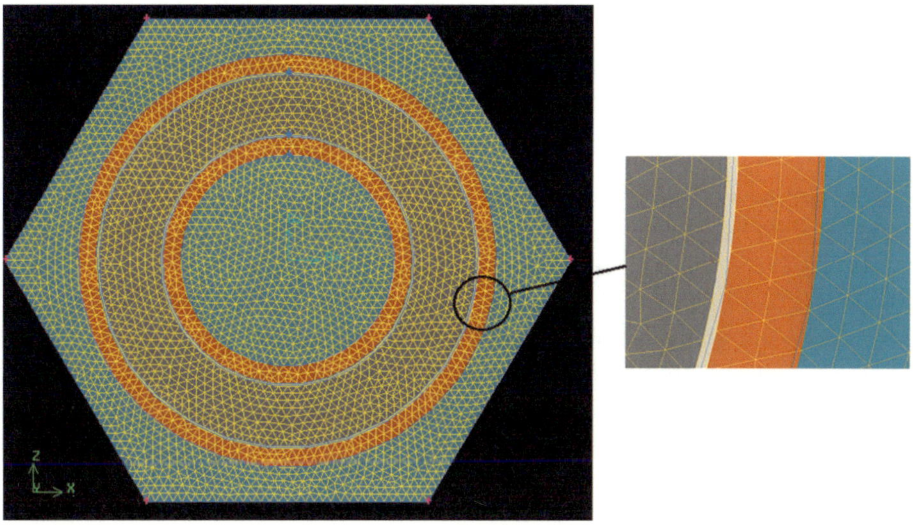

Figure 4.
The quality of mesh in the fuel rod.

$$\nabla.\left(\rho_{nf}VV\right) = -\nabla p + \nabla.\tau \tag{2}$$

$$\nabla.\left(\rho_{nf}Ve\right) = -\nabla.\left(k_{nf}\nabla T\right) - \nabla.(pV) + \nabla.(\tau.V) \tag{3}$$

Where ρ is the density of the nanofluid, V is the velocity vector, p is the pressure, τ is the stress tensor, e is the specific total energy, and T is the temperature. k_{nf} is the conductivity of the nanofluid.

The CFD is used to solve the denoted Navier–Stokes equations. In this code, the SIMPLEC (Semi-Implicit Method for Pressure Linked Equations-Consistent) algorithm, a modified form of the SIMPLE algorithm, is used for numerical procedures in CFD. The algorithm manipulates the same steps as the SIMPLE algorithm with a minor change in the momentum equations, allowing the SIMPLEC velocity correction equations to delete fewer important expressions than those deleted in SIMPLE. It tries to avoid the effects of reducing dropping velocity neighbor correction terms.

3. Thermophysical properties of nanofluid

In this study, it simulated the core of the VVER-1000 reactor with solid and annular fuel using Al_2O_3 nanofluid as coolant. **Table 3** shows the physical and thermal properties of this nanoparticle and base fluid [29].

The correlations for density (ρ_{nf}), specific heat $\left(Cp_{nf}\right)$, viscosity (μ_{nf}), and thermal conductivity (k_{nf}) of the nanofluid can be obtained [28, 30, 31]:

$$\rho_{\mathrm{nf}} = (1-\varphi)\rho_{\mathrm{nf}} + \varphi\rho_{\mathrm{bf}} \tag{4}$$

$$\mathrm{Cp}_{\mathrm{nf}} = (1-\varphi)\mathrm{Cp}_{\mathrm{bf}} + \varphi\mathrm{Cp}_{\mathrm{bf}} \tag{5}$$

$$\mu_{\mathrm{nf}}/\mu_{\mathrm{bf}} = 123\varphi^2 + 7.3\varphi + 1 \tag{6}$$

Property	C (J/kg K)	$\rho(kg/m^3)$	k (W/mK)	$\alpha(m^2/s)$
water	4179	997.1	0.605	1.47
Al_2O_3	765	3970	40	1317

Table 3.
Thermo-physical properties of nanoparticles and base fluids [16].

$$k_{nf}/k_{bf} = 4.97\varphi^2 + 2.72\varphi + 1 \tag{7}$$

In this simulation, φ is the volume fraction of nanofluid considered 5,10, 20.
ρ_{bf}, Cp_{bf}, μ_{bf}, and k_{bf} are density, specific heat, viscosity, and thermal conductivity of the pure fluid, respectively.

4. Results and discussion

4.1 Validation

The simulation results are presented for a typical VVER-1000 reactor with the specifications listed in **Table 1**. To evaluate simulation accuracy, the obtained results for solid fuel are compared with three other studies. Firstly, the distribution of the axial coolant temperature with 5 and 10 volume fractions of nanofluid is presented in **Figure 5**.

The obtained results are compared with the porous media approaches solved by Zarifi et al. [25]. In the research of Zarifi, the conservation equations are solved by numerical methods using visual FORTRAN language; Furthermore, the nanofluids analysis was compared with an analysis of pure water. Finally, the applied approaches in the Zarifi study were validated using the COBRA-EN code for pure water [25].

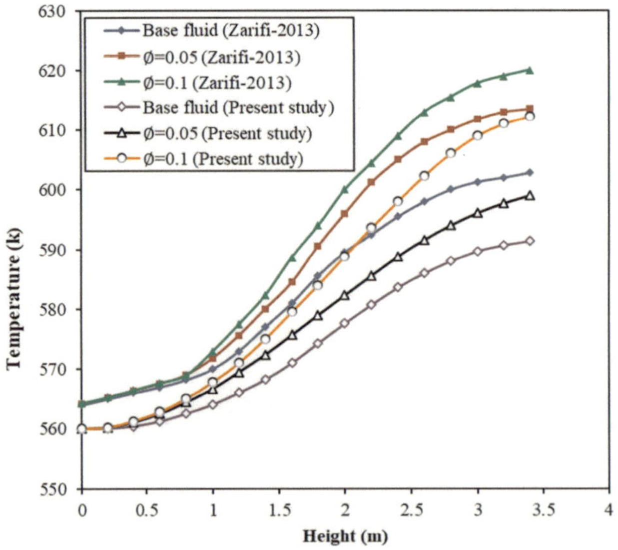

Figure 5.
Axial coolant temperature distribution in the solid fuel [13].

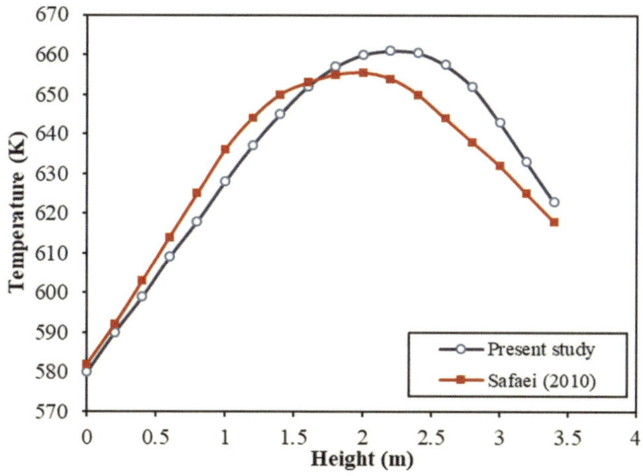

Figure 6.
Axial clad temperature distribution in the solid fuel [21].

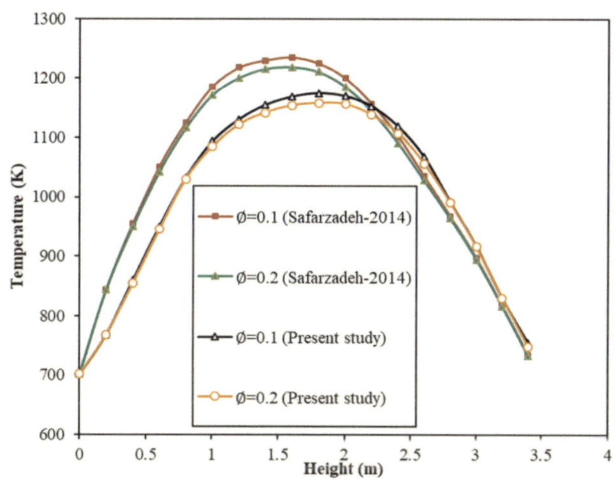

Figure 7.
Axial fuel temperature distribution in the solid fuel [22].

Second, the distribution of the axial clad temperature in solid fuel with pure water is compared with the COBRA-EN code that Safaei has used (**Figure 6**) (In the simulation of Safaei, the COBRA-EN code is improved to make a thermal–hydraulic analysis for the VVER reactor. To validate it, this calculation is compared with reactor FSAR and analytical approaches.) [32].

Third, **Figure 7** presents the axial fuel temperature with 10 and 20 volume fractions of nanofluid. It is compared with the DRAGON/DONJON code used by Safarzadeh (In the simulation of Safarzadeh, the DRAGON, DONJON, and a thermal–hydraulic model are used for the coupled analysis of the nanofluid core. The results are compared with the final safety analysis report (FSAR).) [33].

Finally, as shown in **Figure 8**, the axial fuel temperature with pure water is demonstrated. It is evaluated with the plant's final safety analysis report (FSAR) [26].

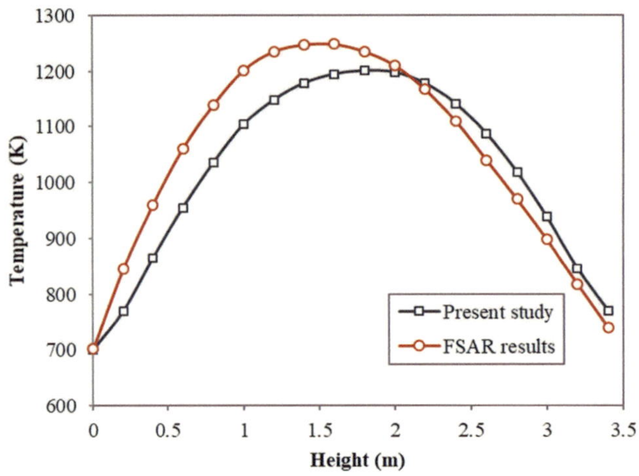

Figure 8.
Axial fuel temperature distribution in the solid fuel [15].

The comparison of the obtained results showed the calculations were in good agreement with other studies. Consequently, the accuracy of the validation is acceptable.

The results showed that due to the increase in heat transfer coefficient by increasing the concentration of Al_2O_3nanofluid, the coolant temperature increased, and the central fuel temperature decreased. It is observed that for 10% by volume of nanoparticles, the difference between the cooling temperature and pure water was about 21°C and the difference between the fuel temperature was about 26°C. This study describes the effect of nanofluid on the cooling system's performance in the reactor core.

4.2 Results

Here, it is simulated annular fuel with pure water. In annular fuel, heat is transferred from internal and external surfaces. **Figures 9** and **10** present the distribution

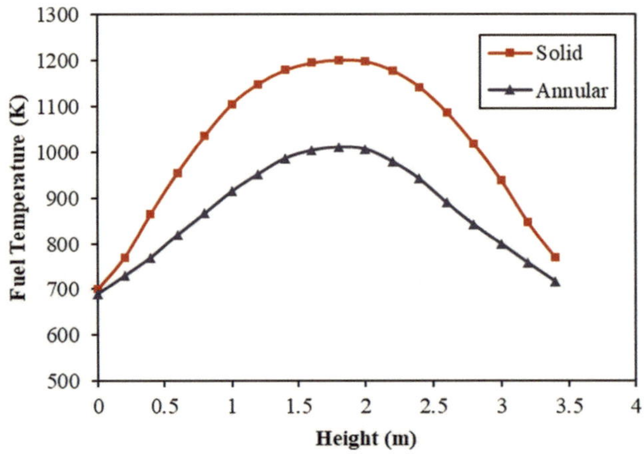

Figure 9.
Comparison of axial fuel temperature distribution.

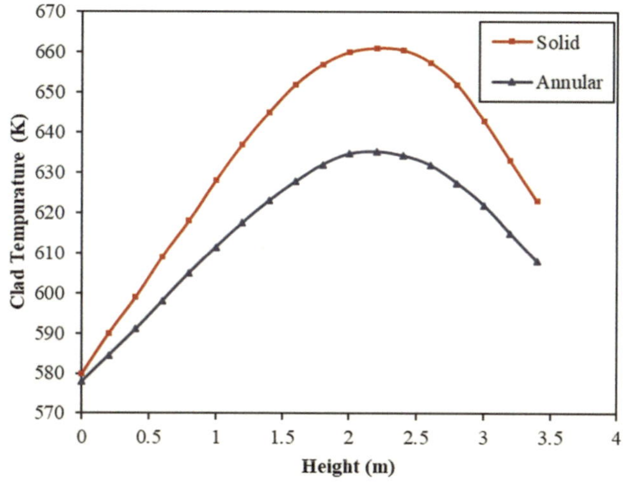

Figure 10.
Comparison of axial clad temperature distribution.

of the fuel and clad temperature in annular fuel and solid fuel with pure water. As can be seen, the distribution of temperature would be cosine shaped, which is due to the cosine shape of the heat generated in the axial direction of the fuel rod.

As can be observed in **Figure 9**, the maximum value of the fuel temperature in solid and annular fuel is 1201 K and 1012 K, respectively. As observed, an approximately 16% reduction in peak temperature at the fuel center was due to the use of annular fuel instead of solid fuel. As shown in **Figure 10**, there is a difference of about 26 degrees between the clad temperature in the annular fuel and the solid fuel.

Furthermore, it simulated the annular fuel with nanofluid. Then, it is compared with pure water. **Figures 11** and **12** compare the distribution of the fuel and clad temperature in annular fuel with 10 and 20 volume nanofluid fractions with pure water.

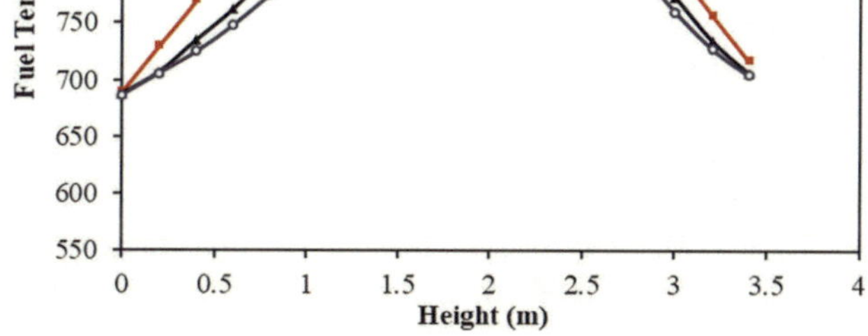

Figure 11.
Comparison of axial fuel temperature distribution in the annular fuel.

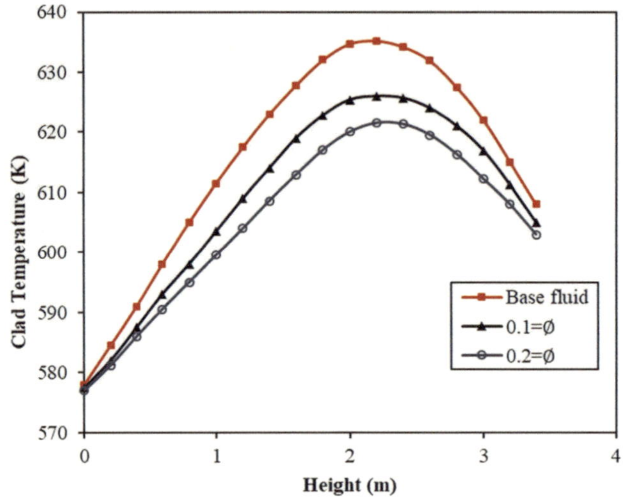

Figure 12.
Comparison of axial clad temperature distribution in the annular fuel.

As can be seen in **Figure 11**, the maximum value of the fuel temperature in annular fuel with pure water is 1012 K, whereas the maximum value of the fuel temperature in annular fuel with 10 and 20 volume percentages of nanoparticles is 920 K and 885 K, respectively. It is concluded that by increasing the concentration of the nanofluid from $\varphi = 0.1$ to$\varphi = 0.2$, the decreasing peak temperature in the centre of the fuel changes from 9–13%. Also, **Figure 12** illustrates that there is about a 13 degrees reduction in the clad temperature with nanofluid compared to the clad temperature with pure water. It was also found that the clad temperature gets closer to the coolant temperature when the nanofluid concentration increases.

In annular fuel, the maximum fuel temperature is reduced because the heat is removed from both sides of the fuel rod. Furthermore, as the heat transfer surface increases, the value of heat flux decreases; thus, it increases the minimum departure from the nucleate boiling ratio (MDNBR) margin. Finally, by increasing the use of the annular fuel, the critical heat flux can be increased.

In the steady-state operational condition, the MDNBR is one of the key limitations of thermal–hydraulic safety. DNBR is calculated from the relation,

$$\mathrm{DNBR} = \frac{q''_{\mathrm{CHF}}}{q''_{\mathrm{actual}}} \tag{8}$$

The nucleate boiling heat flux cannot be increased indefinitely, and it is called the critical heat flux (CHF) at some value. The reactor core must be designed to keep the DNBR larger than the minimum allowable value during steady-state operation, normal operational transients, and anticipated operational occurrences. In annular fuel, the minimum value of DNBR is 1.97, which is higher than the 1.75 value reported by the FSAR for the typical VVER-1000. In other words, the maximum actual heat flux in the annular fuel is 1389.93 kW/m^2, while in FSAR, this value is equal to 1570 kW/m^2 [15].

According to the description, it was found that by using the nanofluid as a coolant, the heat generated in the core from the fission reaction in the fuel could be further

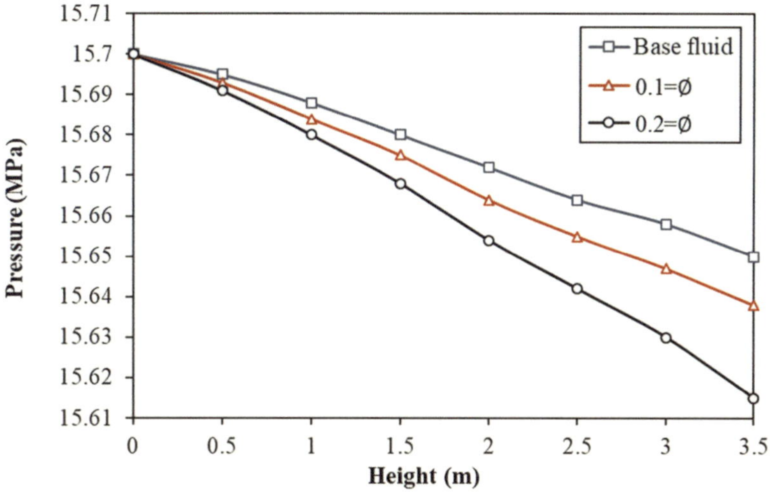

Figure 13.
Comparison of the core pressure drop in the annular fuel.

transferred by the nanofluid. Therefore, safety margins are improved for various transient accidents and crashes.

As observed in **Figure 13** for the annular fuel, the pressure drop changes along the channel in different concentrations of nanoparticles. It is explained that as the concentration of nanofluids increases, the pressure drop increases.

5. Conclusions

This work studied the effect of the using annular fuel rods on thermal–hydraulic characteristics of a typical VVER-1000 reactor. The results of the annular fuel showed that the increase in the heated surface improves the margin from the peak temperature of the fuel to melting. The use of Al_2O_3 nanofluid as a coolant in the VVER-1000 nuclear reactor is also recommended. The results showed that by improving the heat transfer coefficient of Al_2O_3 nanofluid to increase the concentration of Al_2O_3 nanoparticles, the coolant temperature increases, and the central fuel temperature decreases. To explain more, there is about 127 K reduction in the maximum value of fuel temperature in the annular fuel with 20 volume percentages of nanoparticles compared to pure water. Furthermore, there is about 13 degrees reduction in the clad temperature with nanofluid compared to the clad temperature with pure water. The results showed the minimum value of DNBR for annular fuel was 1.97. Finally, it is determined that the use of nanofluid with annular fuel rods has the best efficiency in a nuclear reactor. By increasing the heated surfaces in the annular fuel and due to the higher heat transfer coefficient of nanofluids compared to pure water, the coolant flow rate can be reduced. As a result, the reactor core can be more compact, and the plant's capital cost is reduced.

Author details

Valiyollah Ghazanfari
Nuclear Fuel Cycle Research School, Nuclear Science and Technology Research Institute, Tehran, Iran

*Address all correspondence to: vghazanfary@aeoi.org.ir

References

[1] Wang Z, Wu Z, Han F, et al. Experimental comparative evaluation of a graphene nanofluid coolant in miniature plate heat exchanger. International Journal of Thermal Sciences. 2018;**130**:148-156

[2] Kim HJ, Lee SH, Lee JH, et al. Effect of particle shape on suspension stability and thermal conductivities of water-based bohemite alumina nanofluids. Energy. 2015;**90**:1290-1297

[3] Gupta M, Singh V, Kumar R, et al. A review on thermophysical properties of nanofluids and heat transfer applications. Renewable and Sustainable Energy Reviews. 2017;**74**:638-670

[4] Li CH, Peterson GP. Experimental studies of natural convection heat transfer of Al2O3/DI water nanoparticle suspensions (nanofluids). Advances in Mechanical Engineering. 2010; **2**:1-10

[5] Das PK, Islam N, Santra AK, et al. Experimental investigation of thermophysical properties of Al2O3-water nanofluid: Role of surfactants. Journal of Molecular Liquids. 2017;**37**: 304-312

[6] Pak BC, Cho YI. 1998. Hydrodynamic and heat transfer study of dispersed fluids with submicron metallic oxide particles. Experimental Heat Transfer. 1998;**11**:151-170

[7] Agarwal R, Verma K, Agrawal NK, et al. Sensitivity of thermal conductivity for Al_2O_3 nanofluids. Experimental Thermal and Fluid Science. 2017;**80**: 19-26

[8] Kiruba R, Vinod S, Zaibudeen AW, et al. Stability and rheological properties of hybrid g-Al2O3 nanofluids with cationic polyelectrolyte additives. Colloids and Surfaces A. 2018;**555**: 63-71

[9] Kumar N, Sonawane SS, Sonawane SH. Experimental study of thermal conductivity, heat transfer and friction factor of Al_2O_3 based nanofluid. International Communications in Heat and Mass. 2018;**90**:1-10

[10] Ghoshdastidar PS. Computer Simulation of Flow & Heat Transfer. New Delhi: Tata McGraw Hill. Edt; 1998

[11] Gupta GP. Rajendra Prakash. Engineering Heat Transfer. New Delhi: NemChand & Brothers 6th Edt. 1994

[12] Eskandari MR, Bavandi A, Mihandoost A. Studies on nuclear fuel rod thermal performance. Energy Procedia. 2012;**14**:142-147

[13] Kazimi MS. Introduction to the annular fuel special issue. Nuclear Technology. Taylor & Francis. 2007;**160**:1-1

[14] Kazimi MS. High performance fuel design for next generation PWRs: Final report. In: MIT-NFC-PR-082. Massachusetts Institute of Technology. Center for Advanced Nuclear Energy Systems. Nuclear Fuel Cycle Program, MIT; 2006

[15] Feng D, Hejzlar P, Kazimi MS. Thermal-hydraulic design of high power density annular fuel in PWRs. Nuclear Technology. 2007;**160**:16-44

[16] Shin CH, Chun TH, Oh DS, In WK. Thermal-hydraulic performance assessment of dual-cooled annular fuel. Nuclear Engineering and Design. 2012; **243**:291-300

[17] Yang YS, Shin CH, Chun TH, Song KW. Evaluation of a dual-cooled annular fuel heat Split and temperature distribution. Journal of Nuclear Science and Technology. 2012; **46**:305-353

[18] Yang YS, Bae KM, Shin CH, Chun TH, Bang JG, Song KW. Conceptual design of OPR-1000 compatible annular fuel assembly. In: Proceedings of ICAPP, Nice, France, May 13–18. 2007. p. 7445

[19] ErfaniNia A, Faghihi F, Hadad K. Prompt and power reactivity coefficients for the next generation VVER-1000 reactor including hexagonal assemblies andannular fuels. Progress in Nuclear Energy. 2012;**61**:41-47

[20] Mozafari M, Faghihi F. Design of annular fuels for a typical VVER-1000 core: Neutronic investigation, pitch optimization and MDNBR calculation. Annals of Nuclear Energy. 2013;**60**: 226-234

[21] Zarifi E, Jahanfarnia G, Veysi F. Subchannel analysis of nanofluids application toVVER-1000 reactor. Chemical Engineering Research and Design. 2013;**91**:625-632

[22] Wu C, Cho TJ, Xu J, Lee D, Yang B, Zachariah MR. Effect of nanoparticle clustering on the effective thermal conductivity of concentrated silica colloids. Physical Review E. 2010;**81**: 011406

[23] Xuan Y, Roetzel W. Conceptions for heat transfer correlation of nanofluids. International Journal of Heat and Mass Transfer. 2000;**43**:3701-3707

[24] Albadr J, Tayal S, Alasadi M. Heat transfer through heat exchanger using Al_2O_3 nanofluid at different concentrations. International Journal of Multiphase Flow. 2013;**35**:427-438

[25] Zarifi E, Jahanfarnia G, Veysi F. Thermal–hydraulic modeling of nanofluids as the coolant in VVER-1000 reactor core by the porous media approach. Annals of Nuclear Energy. 2013;**51**:203-212

[26] FSAR of BNPP-1, 2003. Final Safety Analysis Report of BNPP-1, 2003.49. BU.10.0.00. FSAR. RDR001

[27] Bianco V, Chiacchio F, Manca O, Nardini S. Numerical investigation of nanofluids forced convection in circular tubes. Application Thermal Engineering. 2009;**29**:3632-3642

[28] Velagapudi V, Konijeti RK, Aduru CS. Empirical correlation to predict thermophysical and heat transfer characteristics of nanofluids. Thermal Science. 2008;**12**:27-37

[29] Hadad K, Rahimian A, Nematollahi MR. Numerical study of single and two-phase models of water/ Al2O3 nanofluid turbulent forced convection flow in VVER-1000 nuclear reactor. Annals of Nuclear Energy. 2013; **60**:287-294

[30] Maïga SEB, Nguyen CT, Galanis N, Roy G. Heat transfer behaviors of nanofluids in a uniformly heated tube. Superlattices and Microstructures. 2004; **35**:543-557

[31] Pak BC, Cho YI. Hydrodynamic and heat transfer study of dispersed fluids with submicron metallic oxide particles. Experimental Heat Transfer. 1998;**11**: 151-170

[32] SafaeiArshi S, Mirvakili SM, Faghihi F. Modified COBRA-EN code to investigate thermal-hydraulic analysis of

the IranianVVER-1000 core. Progress in Nuclear Energy. 2010;**52**:589-595

[33] Safarzadeh O, Shirani AS, Minuchehr F, Saadatian-derakhshandeh F. Coupled neutronic/thermo-hydraulic analysis of water/Al2O3 nanofluidsin a VVER-1000 reactor. Annals of Nuclear Energy. 2014;**65**:72-77

Chapter 6

CFD Applications in Ground Source Heat Pump System

Yajiao Liu, Junhu Dang, Xiaosen Dai, Hao Luo, Yipeng Wu and Tiecheng Zhao

Abstract

In ground source heat pump (GSHP) system, computational fluid dynamics (CFD) is commonly used to conduct simulation analysis of its operating characteristics. Particularly, ground heat exchanger (GHE) is the most core component of GSHP system, and the heat transfer characteristics of which with soil around will directly affect the efficiency of the entire system. Thus, CFD is always applied to predict the process of heat transfer around GHE and its influence on heat exchange process. In this chapter, a 3-D numerical model considering dynamic surface condition and initial soil temperature distribution is developed to investigate the thermal performance of helix ground heat exchanger (HGHE) on basis of CFD, and the main influencing factor (inlet water temperature) is studied with the established model. In addition, the experimental investigation is carried out to verify the accuracy of the model. The results are of great significance for exploring the application of CFD in GSHP system.

Keywords: CFD, ground source heat pump, heat exchanger, thermal performance, geothermal energy

1. Introduction

Shallow geothermal energy resource is a kind of renewable resources, mainly referring to the low-temperature heat energy, also known as geothermal energy, in the earth's shallow surface within hundreds of meters (less than 200 m). Different from the traditional deep geothermal energy, it refers to the stratum thermal energy in the general layer, the temperature of which is less than 25°C, and its energy mainly comes from solar radiation and earth gradient warming. Compared with deep geothermal energy in the traditional sense, shallow geothermal energy is not affected by geological factors and generally exists in the surface of the earth, with many advantages such as wide distribution and renewable [1–3].

Due to the low temperature of shallow geothermal energy, which is not easy to be extracted, it was not used by people in the early years. With the progress of science and technology and the attention paid to the impact of natural environment, as a renewable, clean, and huge kind of energy, it has gradually been widely paid attention to. At present, the shallow geothermal energy is mainly used in the air conditioning system of buildings. The method is to extract the low-grade underground shallow

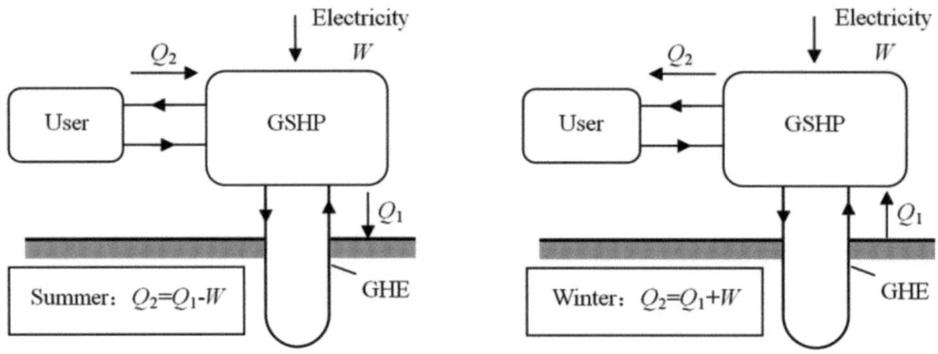

Figure 1.
Schematic of ground source heat pump system.

geothermal energy through the GSHP system and make use of it. **Figure 1** shows the schematic diagram of the GSHP system. By burying the GHE underground, it makes heat exchange with the relatively constant temperature stratum. In winter, it takes the earth as the high-temperature potential heat source and uses a small amount of high-grade energy (electric energy) to extract heat from the earth and provides heat for the user. On the contrary, in summer, it takes the earth as the low-temperature potential heat source and releases the excess heat of the user into the earth [4–8].

Traditional GSHP system can be divided into two kinds: water source heat pump technology and soil source heat pump technology. The heat source of the water source heat pump system is the underground or surface water, and the vast majority of the water source heat pump system directly extracts underground or surface water, which is an open system. The soil source heat pump system is a closed system by burying the GHE in the ground and exchanging heat with the surrounding rock and soil through the flow of the working medium inside the GHE. Because the underground or surface water used by water source heat pump system contains microorganisms, calcium and magnesium ions, and sulfate ions, the scale and blockage always occur in the system, resulting in poor system stability. At the same time, it is also restricted by hydrogeological conditions; thus, for the sake of realizing the development and utilization of shallow geothermal energy resources, the soil source heat pump system is widely used in the project at present.

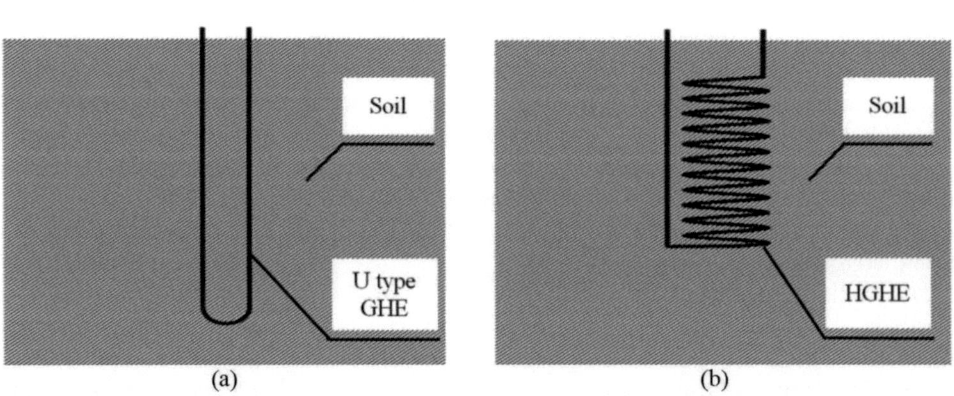

Figure 2.
Schematic of ground source heat pump system. (a) Traditional U-type GHE, (b) New-type HGHE.

Buried GHE is the most core component in the GSHP system. As shown in **Figure 2a**, the application research of the traditional U-type GHE started earlier and developed more mature. Therefore, the U type GHE is the most widely used, and most GSHP projects adopt this type of buried pipe. Compared with the U type GHE, the HGHE is a new type of geothermal exchanger, as shown in **Figure 2b**. Because the spiral pipe is arranged along the cylindrical wall, its heat transfer area and heat transfer amount are large, with low initial investment, high heat transfer efficiency, and other significant advantages, and HGHE with its unique advantages has been rapidly developed in recent years, effectively solving the economic and technical problems of traditional U-type GHE.

Due to the complicated hydrogeological conditions involved in the actual engineering of GHE, it is difficult to obtain the heat transfer process of buried pipe by experimental means due to the limitation of the region and testing conditions. For this reason, scholars around the world have carried out a lot of numerical heat transfer simulation works on GHE, mainly by using CFD technology to achieve the simulation research on the heat transfer characteristics and heat transfer process of buried pipes, for the sake of maximizing the heat transfer efficiency and improving the overall efficiency of the GSHP system. Angelo Zarrella [9–11] et al. established a numerical heat transfer model based on U-type and helix GHEs by using the electrical resistance analysis method. They considered the influence of external environment on soil boundary conditions and conducted experimental research. The model mainly considers the convective heat transfer of helical fluid, and the radial and axial heat conduction of soil and pile foundation, and ignores the heat transfer along the Angle direction. Salsuwanda Selamat [12] aiming at the design optimization problem of horizontal HGHE, established a three-dimensional numerical model and used CFD software for simulation calculation to simulate and optimize each parameter. Gyu-Hyun Go et al. [13] established a three-dimensional numerical model of horizontal HGHE and simulated its heat transfer characteristics. The accuracy of the model was verified through indoor thermal response experiment. Qiang Zhao et al. [14] established a three-dimensional unsteady heat transfer model reflecting the heat transfer process of fluid and soil in the tube for the GHE and used the finite element method to solve and calculate. Then, the heat transfer characteristics of U-type, W-type, and helix GHEs are simulated and calculated, respectively. The research shows that the heat transfer efficiency of HGHE is higher than that of other two types of heat exchangers.

In order to further elaborate and introduce the application of CFD technology in the GSHP system, this chapter takes the new-type HGHE as an example and establishes its three-dimensional numerical model, which fully considers the external dynamic environmental conditions and initial soil temperature conditions. Then, the ANSYS_FLUENT software, which is commonly used in CFD technology, is used to simulate and study the heat transfer characteristics of HGHE under different inlet water temperature conditions. Finally, the accuracy of the numerical model is verified by experimental means.

2. Description of numerical model

2.1 Physical model

The research object of this chapter is made of polyethylene (PE) pipe, which is applied in Chongqing, China. The construction process is strictly in accordance with the relevant requirements of *Technical Specifications for Ground Source Heat Pump System Engineering*. And the specific parameters are shown in **Table 1**.

Parameters	Spiral pitch in the depth direction/m	Diameter/m	Height/m	Depth of the top of HGHE from ground surface/m	Inside diameter of coil pipe/m
HGHE	0.08	1.14	2.00	1.00	0.02

Table 1.
Main structural parameters of HGHE.

Using the Design Modeler module in ANSYS_FLUENT, which is a professional finite element analysis tool, this work established the geometric model of HGHE. According to the real size, the surrounding soil was set as a cylinder in the physical model. And the geometric model is shown in **Figure 3a**.

Then, the meshing module of ANSYS_FLUENT software is used to divide the grid of the built geometric model of HGHE. The grid division follows the following principles: If the temperature field or velocity field in a certain position and direction has significant fluctuations, the area should be divided into dense grids. On the contrary, if the fluctuation of temperature field or velocity field is not obvious, the meshing of this area can be loosened appropriately. Because the scale of fluid flow and heat transfer in the spiral buried pipe are small and fluctuate greatly during heat transfer, the grid of the spiral buried pipe and its nearby soil area are encrypted, as shown in **Figure 3b**. The specific grid division is as follows:

1. Hexahedral mesh is selected for mesh division of HGHE. Hexahedral mesh can reduce the number of elements and improve the convergence speed and analysis accuracy. Because the flow direction of the circulating medium in the tube changes at all times, the flow field fluctuates significantly, and local encryption is carried out during grid division to avoid large inclination angle of the grid in this area, which will promote the convergence speed and the accuracy of the calculation results.

2. The soil around the HGHE is hexahedral mesh, in which the area adjacent to the buried pipe is locally encrypted to avoid large inclination angle of the grid in this area, which will also promote the convergence speed and the accuracy of the calculation results.

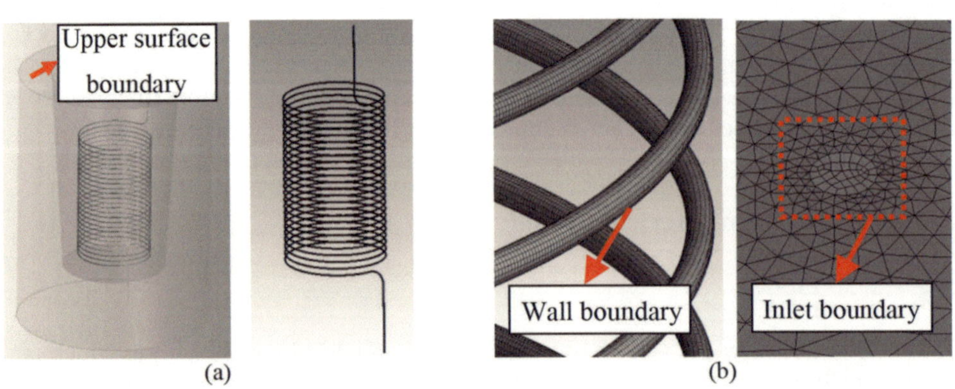

Figure 3.
Mesh division of physical model of HGHE. (a) Physical model of HGHE. (b) Mesh division of HGHE.

3. Mesh with higher quality can accelerate the convergence speed, reduce the error in solving, avoid the divergence of numerical simulation results, and ensure the accuracy of results. ANSYS_FLUENT software is required to ensure that the minimum orthogonal quality is greater than 0.1 or the maximum inclination is less than 0.95. Most of the grids in the model established in this chapter have good quality, and a few grids with poor quality are in the acceptable range of ANSYS_FLUENT. Therefore, the grids used for numerical calculation can meet the requirements of ANSYS_FLUENT.

2.2 Model assumption

To simplify the solution, the following assumptions are made for the numerical model:

1. The soil around the buried pipe is regarded as an idealized uniform porous medium, and the thermal physical parameters of the soil remain constant in the heat transfer process.

2. The solid skeleton of the soil has no deformation, and the inner space is interconnected, with uniform porosity.

3. The contact thermal resistance between buried pipe and soil is not considered.

4. The influence of soil seepage and water migration is not considered.

5. Ignoring the influence of the HGHE around, the adiabatic boundary conditions were assumed around the cylindrical soil.

6. The local transient thermal equilibrium state is considered between soil and buried pipe.

2.3 Governing equations

The soil around the HGHE is regarded as a uniform porous medium, while the flow state of the medium in the spiral buried pipe is turbulent. Therefore, the governing equations of fluid flow and heat transfer in pipe and heat transfer of soil are established respectively.

2.3.1 Governing equations of fluid flow and heat transfer in pipe

Due to the turbulent flow in HGHE, an appropriate turbulence model should be selected for calculation. The $k - \varepsilon$ equations are the most widely used in numerical computation and engineering applications. They include:

Turbulent kinetic energy equation:

$$\frac{\partial(\rho k)}{\partial \tau} + \frac{\partial(\rho k u_i)}{\partial x_i} = \frac{\partial}{\partial x_j}\left[\left(\mu + \frac{\mu_i}{\sigma_k}\right)\frac{\partial k}{\partial x_j}\right] + G_k + G_b - \rho\varepsilon - Y_M + S_k \tag{1}$$

Dissipation rate equation:

$$\frac{\partial(\rho\varepsilon)}{\partial\tau}+\frac{\partial(\rho\varepsilon u_i)}{\partial x_i}=\frac{\partial}{\partial x_j}\left[\left(\mu+\frac{\mu_i}{\sigma_\varepsilon}\right)\frac{\partial\varepsilon}{\partial x_j}\right]+C_{1\varepsilon}\frac{\varepsilon}{k}(G_k+C_{3\varepsilon}G_b)-C_{2\varepsilon}\rho\frac{\varepsilon^2}{k}+S_\varepsilon \tag{2}$$

where ρ is the fluid density, kg/m^3. τ is the time, s. k, ε respectively represents the dissipation rate of turbulent kinetic energy and pulsation energy. u_i, u_j is the hourly average speed, m/s. μ, μ_i respectively represent the dynamic viscosity and turbulent viscosity. G_k is the turbulent kinetic energy generation term caused by average velocity gradient. G_b is the turbulent kinetic energy generation term. Y_M is the contribution of pulsation expansion. $C_{1\varepsilon}$, $C_{2\varepsilon}$, and $C_{3\varepsilon}$ are the empirical constants, the values of which are 1.44, 1.92, and 0.09, respectively. σ_k, σ_ε are the Pr number, and the values are 1 and 1.3, respectively. S_k, S_ε are the user-defined source entries.

The circulating fluid in the buried pipe is water, which is an incompressible fluid and ignores the action of gravity. Thus, it can be simplified as:

Turbulent kinetic energy equation:

$$\frac{\partial(\rho k)}{\partial t}+\frac{\partial(\rho k u_i)}{\partial x_i}=\frac{\partial}{\partial x_j}\left[\left(\mu+\frac{\mu_i}{\sigma_k}\right)\frac{\partial k}{\partial x_j}\right]+G_k-\rho\varepsilon \tag{3}$$

Dissipation rate equation:

$$\frac{\partial(\rho\varepsilon)}{\partial t}+\frac{\partial(\rho\varepsilon u_i)}{\partial x_i}=\frac{\partial}{\partial x_j}\left[\left(\mu+\frac{\mu_i}{\sigma_\varepsilon}\right)\frac{\partial\varepsilon}{\partial x_j}\right]+\frac{\varepsilon}{k}(C_{1\varepsilon}G_k-C_{2\varepsilon}\rho\varepsilon) \tag{4}$$

Among them, the Prandtl number, turbulent kinetic energy k, and dissipation rate ε can be automatically calculated by FLUENT software through fluid physical parameters and velocity.

The standard $k-\varepsilon$ model, mass conservation equation, momentum conservation equation, and energy conservation equation together constitute the governing equation of flow and heat transfer, the general form of which is:

$$\frac{\partial(\rho\phi)}{\partial\tau}+\mathrm{div}(\rho U\phi)=\mathrm{div}(\Gamma_\phi\mathrm{grad}\phi)+S_\phi \tag{5}$$

where ϕ is the universal variable. Γ_ϕ is the generalized diffusion coefficient. S_ϕ is the generalized source term.

The values of parameters in the general equation are shown in **Table** 2:

Equation	ϕ	Γ_ϕ	S_ϕ
Turbulent kinetic energy equation	k	$\mu+\mu_i/\sigma_k$	$G_k-\rho\varepsilon$
Dissipation rate equation	ε	$\mu+\mu_i/\sigma_\varepsilon$	$\varepsilon(C_{1\varepsilon}G_k-C_{2\varepsilon}\rho\varepsilon)/k$
Mass conservation equation	1	0	0
X- momentum conservation equation	u	$\eta=\mu+\mu_i$	$-\frac{\partial p}{\partial x}+\frac{\partial}{\partial x}\left(\eta\frac{\partial u}{\partial x}\right)+\frac{\partial}{\partial y}\left(\eta\frac{\partial v}{\partial x}\right)+\frac{\partial}{\partial z}\left(\eta\frac{\partial w}{\partial x}\right)$
Y-momentum conservation equation	v	$\eta=\mu+\mu_i$	$-\frac{\partial p}{\partial y}+\frac{\partial}{\partial x}\left(\eta\frac{\partial u}{\partial y}\right)+\frac{\partial}{\partial y}\left(\eta\frac{\partial v}{\partial y}\right)+\frac{\partial}{\partial z}\left(\eta\frac{\partial w}{\partial y}\right)$
Z- momentum conservation equation	w	$\eta=\mu+\mu_i$	$-\frac{\partial p}{\partial z}+\frac{\partial}{\partial x}\left(\eta\frac{\partial u}{\partial z}\right)+\frac{\partial}{\partial y}\left(\eta\frac{\partial v}{\partial z}\right)+\frac{\partial}{\partial z}\left(\eta\frac{\partial w}{\partial z}\right)$
Energy conservation equation	T	$\mu/Pr+\mu_i/\sigma$	0

Table 2.
The value of each parameter in the general equation.

DOI: http://dx.doi.org/10.5772/intechopen.109574

2.3.2 Governing equations of heat transfer of soil

The heat transfer process of soil around the spiral buried pipe also follows the general governing Eq. (5), and the values of its parameters are shown in **Table 2**.

2.4 Initial and boundary conditions

2.4.1 Upper surface dynamic boundary condition

In this chapter, the data in the *Special Meteorological Data Set for the Analysis of the Thermal Environment of Buildings in China* are adopted to simulate the hourly meteorological data of the typical meteorological year of Chongqing, as shown in **Figure 4**, and the dynamic boundary conditions of the upper surface are written into ANSYS_FLUENT software through user-defined function for simulation.

2.4.2 Initial soil temperature condition

Because HGHE is generally shallow buried depth, shallow soil initial temperature distribution is often uneven due to external influence. Based on this, in order to make the simulation more suitable to the actual project, this chapter selects the soil temperature distribution of the hottest month in Chongqing as the initial temperature condition of the simulation, as shown in **Figure 5**. Also, the initial soil temperature conditions are written into ANSYS_FLUENT software through user-defined function for simulation calculation.

2.4.3 Parameter of soil

The research background of this chapter is in Chongqing, China, and the soil type widely distributed in Chongqing is strongly weathered mudstone, the specific parameters of which are determined by experiment, as shown in **Table 3**.

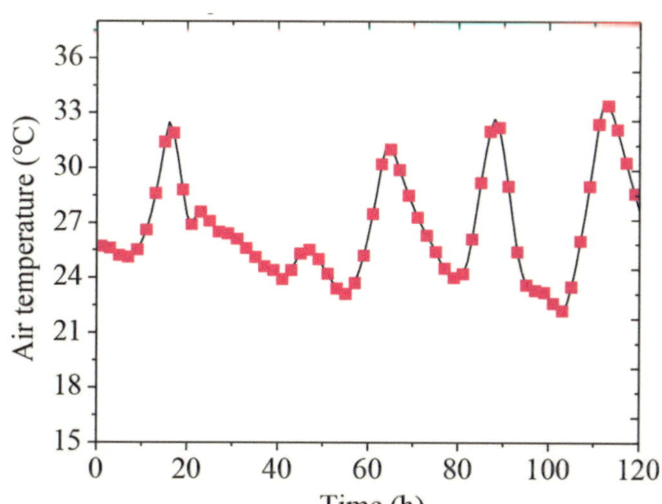

Figure 4.
Hourly meteorological parameters in Chongqing of the typical year.

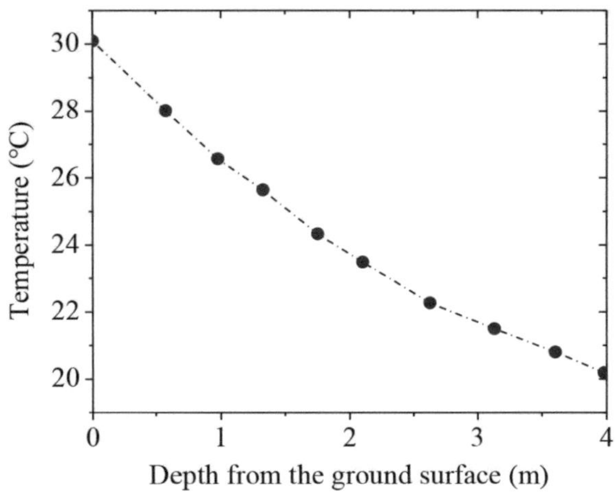

Figure 5.
Initial soil temperature distribution in the depth direction in the hottest month of Chongqing.

Soil type	Strongly weathered mudstone
Density/kg m^{-3}	2503
Specific heat capacity J kg^{-1} °C^{-1}	1085
Porosity	0.3
Comprehensive thermal conductivity/W m^{-1} °C^{-1}	1.24

Table 3.
Thermophysical parameters of strongly weathered mudstone.

2.5 CFD simulation setup

The HGHE model in this chapter is a three-dimensional model, so the option of 3D model should be selected on the initiator interface. In the calculation of the three-dimensional model, ANSYS_FLUENT provides two different calculation methods, and the two different calculation methods have their own suitable situations. The calculation method of single precision is rough, but the amount of calculation is relatively small, and it is applied to the case that the model is regular. The calculation method of double precision is mainly used for the situation that the length and size of the fluid domain is large, the fluid domain has more than one part, and each domain is connected by the small size pipeline with drastic temperature changes and high thermal conductivity. Therefore, this chapter chooses to use the double-precision solver to solve the problem.

This chapter mainly simulates the flow characteristics of fluid in HGHE, which belongs to the range of incompressible flow fluid, so the uncoupled implicit solution is adopted. Because the heat transfer process in the buried pipe was ongoing all the time and the data fluctuated within the range, the transient computing mode was selected. At the same time, the energy equation option is opened, the $k - \varepsilon$ flow model is adopted, and the near-wall function is modified at the same time.

3. Simulation results

The HGHE mainly relies on the temperature difference between the fluid in the pipe and the soil outside the pipe for heat transfer, so the temperature of the fluid in the pipe will directly affect the heat transfer process of the buried pipe, and the temperature of the fluid in the pipe is determined by the inlet water temperature. Therefore, this chapter focuses on studying the influence of different inlet water temperatures on the heat transfer characteristics of the HGHE and the soil thermal response. According to the relevant provisions of the specification *Water (Ground) Source Heat Pump Unit (GB/T 19409-2013)* and combined with the actual project, four different inlet water temperatures (32, 35, 38, and 41°C) were selected at equal intervals to conduct the simulation research. The spiral tube flow rate was 170 L/h, and the continuous operation was simulated for 120 h. The variation characteristics of soil temperature field and heat flux per unit pipe length with time were analyzed.

3.1 Heat transfer capacity

Figure 6 shows the change curve of heat transfer capacity of HGHE with time under different inlet water temperatures. **Figure 6a** shows the change of outlet water temperature of HGHE with running time under four different working conditions. It can be found that outlet water temperature gradually increases with time and finally reaches the stable state. Corresponding **Figure 6b** shows the change of heat flux per unit pipe length with running time under four working conditions. The heat flux per unit pipe length gradually decreases with the passage of time and finally reaches a stable state. Both the outlet temperature curve and the heat flux curve have experienced two stages: the dramatic change stage in the initial stage of operation and the gentle change stage in the later stage. It can be seen that the higher the inlet water temperature is, the more drastic the change at the early stage of operation. This is because the higher the inlet water temperature is, the more heat will accumulate near the spiral pipe in the early stage, which makes the heat exchange resistance of the spiral pipe greater; thus, the heat exchange per unit pipe length decreases more dramatically.

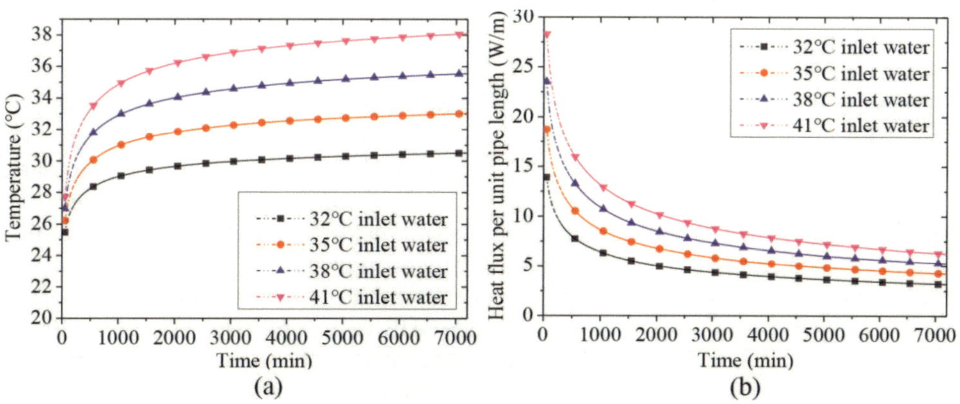

Figure 6.
Heat exchange capacity changing with time of different inlet water temperature conditions. (a) Outlet water temperature changing with time. (b) Heat flux per unit pipe length changing with time.

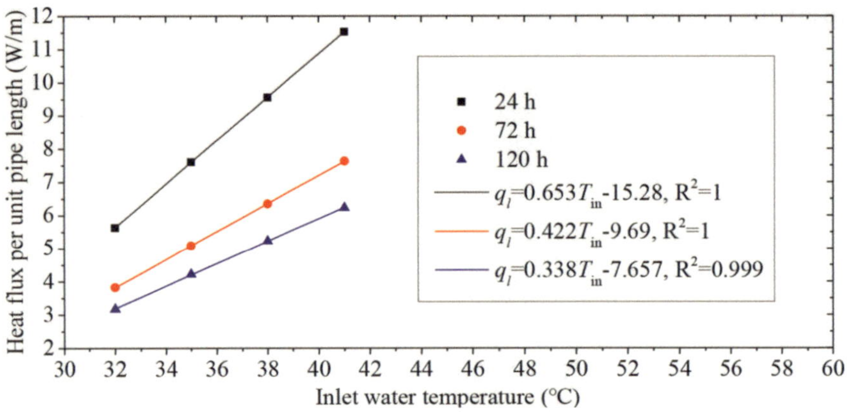

Figure 7.
Heat flux per unit pipe length changing with inlet water temperature.

In order to more clearly explore the influence of inlet water temperature on the heat transfer capacity of HGHE, the relationship curves between heat flux per unit pipe length and inlet water temperature at different times (24, 72, and 120 h) are made, as shown in **Figure 7**. It can be seen that under the three moments, the heat flux per unit pipe length has a linear increasing trend with the increase of the inlet water temperature. When the heat removal operation reaches 24 h, the heat flux per unit pipe length of HGHE is 5.63, 7.60, 9.56, and 11.52 W/m, respectively, for the inlet water temperature of 32, 35, 38, and 41°C. Under the conditions of four different inlet temperatures (32, 35, 38, and 41°C), the heat flux per unit pipe length of HGHE is 3.83, 5.08, 6.35, and 7.63 W/m, respectively, at the moment of 72 h. When the heat removal operation reaches 120 h, the heat flux per unit pipe length of HGHE is 3.18, 4.22, 5.23, and 6.23 W/m for the inlet water temperature of 32, 35, 38, and 41°C, respectively. It is concluded that when the system runs to the late stable state, the linear increase range of heat flux per unit pipe length of HGHE becomes small with the inlet water temperature. For example, at the moment of 120 h, the heat flux per unit pipe length increases only about 0.34W/m when the inlet temperature increases by 1°C.

3.2 Thermal response of soil

Figure 8 shows the temperature distribution field of half longitudinal sections under different inlet water temperature conditions. It can be seen from the figure that the higher the inlet water temperature is, the more obvious the soil heat accumulation phenomenon near the spiral buried pipe is, and the stronger the thermal interference effect is. In particular, the temperature of the soil inside the HGHE increases dramatically with the increase of the inlet water temperature, which is mainly due to the limited volume of soil in the pile. With the progress of the heat removal condition, the heat discharged from the spiral pipe around HGHE will be transferred to the pile, resulting in heat accumulation, while the heat inside HGHE can only be discharged through the upper and lower opening surface. Therefore, the increase of inlet water temperature means that more heat will accumulate inside HGHE, so the thermal interference inside the pile is more and more serious, which will also greatly hinder the improvement of the heat transfer capacity of HGHE. It can also be seen that the

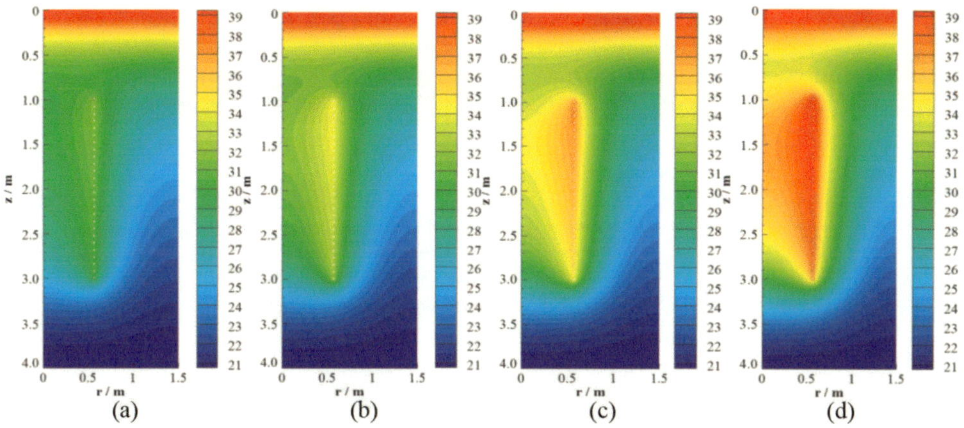

Figure 8.
Soil temperature distribution of different inlet water temperature conditions. (a) 32°C, (b) 35°C, (c) 38°C, (d) 41°C.

increase of inlet water temperature has little influence on the thermal action range of HGHE, but mainly increases the thermal response temperature of soil within the thermal action range. Therefore, it shows that increasing the inlet water temperature of HGHE has no significant effect on improving the heat exchange performance. On the contrary, it will increase the outlet temperature of HGHE, resulting in higher condensing temperature and lower the efficiency of the GSHP system.

4. Model validation

In order to further verify the accuracy of the numerical model established in this chapter, a heat transfer experimental platform corresponding to the HGHE was built, as shown in **Figure 9**. The verification experiment was carried out, and the measured project was compared with the simulated results, through which the verification of the model could be completed.

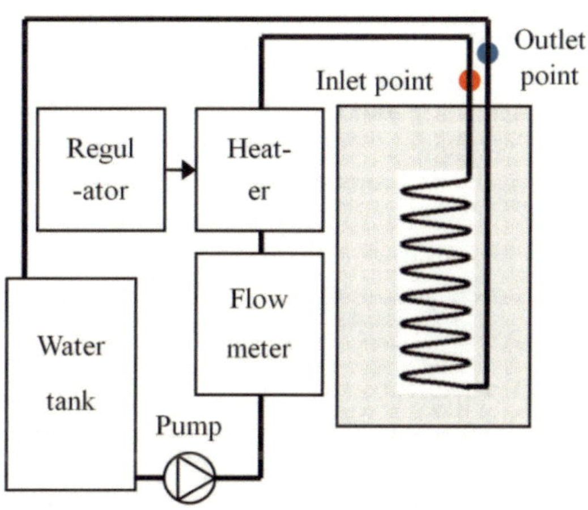

Figure 9.
Heat transfer experimental system of HGHE.

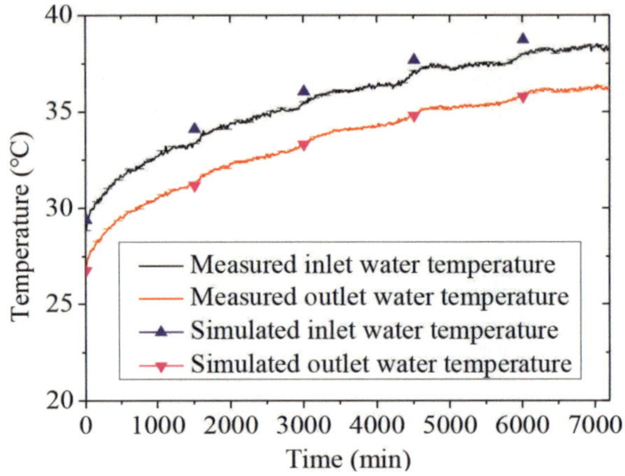

Figure 10.
Comparison of simulation and experiment results.

As can be seen from the comparison curve of inlet and outlet water temperature in **Figure 10**, the numerical simulation results are close to the experimental results, and the maximum difference is less than the 5% relative error commonly used in engineering. But the simulation results are slightly larger than the test results. In practice, there may be a series of conditions such as underground water migration, which are simplified in the model, so there are differences, but the differences are not large, resulting in the measured outlet temperature that is slightly lower than the simulated outlet temperature. In general, the operating results of the model are close to the measured values of the actual project, and the model is basically consistent with the actual values, indicating that the model is basically correct and has a certain guiding role for the actual project.

5. Conclusions

This chapter focuses on the application of CFD technology in GSHP system. Firstly, for the simulation research field of GHE heat transfer characteristics and heat transfer process, the research progress of scholars from various countries is reviewed. Then, in order to further elaborate and introduce the application of CFD technology in GSHP system, the HGHE is taken as an example. Considering the dynamic environment conditions and the initial soil temperature conditions, the ANSYS_FLUENT software commonly used in CFD technology was used to simulate and study the heat transfer characteristics of HGHE under different inlet water temperature conditions, and the accuracy of the numerical model was verified by experimental means. The following conclusions are reached through the research:

1. CFD technology can accurately reflect the actual heat transfer process of HGHE, and it has a certain guiding role for practical engineering.

2. The heat flux per unit pipe length increases linearly with the increase of the inlet water temperature. When the inlet water temperature of HGHE is low, the heat exchange of the buried pipe can be increased by increasing the inlet water

temperature. However, when the inlet water temperature of HGHE is high, continuing to increase the inlet water temperature will seriously aggravate the soil thermal interference near the spiral pipe, which will greatly hinder the heat transfer process of HGHE. At the same time, too high inlet water temperature will increase the outlet water temperature, indirectly causing higher condensation temperature, resulting in lower efficiency of the GSHP system. Therefore, the thermal interference effect and the condensing temperature of GSHP system should be comprehensively considered in the actual project, and the appropriate inlet water temperature of HGHE should be selected.

Acknowledgements

This work received support from Natural Science Foundation of China [52106283].

Conflict of interest

We declare that we have no financial and personal relationships with other people or organizations that can inappropriately influence our work. There is no professional or other personal interest of any nature or kind in any product, service, and/or company that could be construed as influencing the position presented in, or the review of, our work.

Appendices and nomenclature

T	temperature (°C)
x, y, z	three-dimensional direction (m)
u_i, u_j	hourly average speed (m s^{-1})
μ, μ_i	dynamic viscosity and turbulent viscosity
G_k	turbulent kinetic energy generation term caused by average velocity gradient
G_b	turbulent kinetic energy generation term
Y_M	contribution of pulsation expansion
$C_{1\varepsilon}$, $C_{2\varepsilon}$, $C_{3\varepsilon}$	empirical constant, the values of which are 1.44, 1.92 and 0.09 respectively
S_k, S_ε	user-defined source entries
ϕ	universal variable
Γ_ϕ	generalized diffusion coefficient
S_ϕ	generalized source term
Greek symbols	
ρ	density (kg m^{-3})
τ	time (s)
k, ε	dissipation rate of turbulent kinetic energy and pulsation energy
σ_k, σ_ε	Pr number, and the values are 1 and 1.3 respectively

Author details

Yajiao Liu*, Junhu Dang, Xiaosen Dai, Hao Luo, Yipeng Wu and Tiecheng Zhao
National Defense Engineering Institute, Academy of Military Sciences, Wuhan, China

*Address all correspondence to: lyj199164@126.com

References

[1] Cui P, Li X, Man Y, et al. Heat transfer analysis of pile geothermal heat exchangers with spiral coils. Applied Energy. 2011;**88**(11): 4113-4119

[2] Li M, Lai ACK. Heat-source solutions to heat conduction in anisotropic media with application to pile and borehole ground heat exchangers. Applied Energy. 2012;**96**:451-458

[3] Zhang W, Yang H, Lu L, et al. Investigation on heat transfer around buried coils of pile foundation heat exchangers for ground-coupled heat pump applications. International Journal of Heat and Mass Transfer. 2012; **55**(21-22):6023-6031

[4] Wang D, Lu L, Cui P. A novel composite-medium solution for pile geothermal heat exchangers with spiral coils. International Journal of Heat and Mass Transfer. 2016;**93**: 760-769

[5] Ghasemi-Fare O, Basu P. A practical heat transfer model for geothermal piles. Energy and Buildings. 2013;**66**:470-479

[6] Park S, Lee S, Park H, et al. Characteristics of an analytical solution for a spiral coil type ground heat exchanger. Computers and Geotechnics. 2013;**49**:18-24

[7] Leroy A, Bernier M. Development of a novel spiral coil ground heat exchanger model considering axial effects. Applied Thermal Engineering. 2015;**84**:409-419

[8] Suryatriyastuti ME, Mroueh H, Burlon S. Understanding the temperature-induced mechanical behaviour of energy pile foundations. Renewable and Sustainable Energy Reviews. 2012;**16**(5):3344-3354

[9] Zarrella A, De Carli M, Galgaro A. Thermal performance of two types of energy foundation pile: Helical pipe and triple U-tube. Applied Thermal Engineering. 2013;**61**(2):301-310

[10] Zarrella A, De Carli M. Heat transfer analysis of short helical borehole heat exchangers. Applied Energy. 2013;**102**: 1477-1491

[11] Zarrella A, Capozza A, De Carli M. Analysis of short helical and double U-tube borehole heat exchangers: A simulation-based comparison. Applied Energy. 2013;**112**:358-370

[12] Selamat S, Miyara A, Kariya K. Numerical study of horizontal ground heat exchangers for design optimization. Renewable Energy. 2016;**95**:561-573

[13] Go G, Lee S, Yoon S, et al. Optimum design of horizontal ground-coupled heat pump systems using spiral-coil-loop heat exchangers. Applied Energy. 2016; **162**:330-345

[14] Zhao Q, Chen B, Liu F. Study on the thermal performance of several types of energy pile ground heat exchangers: U-shaped, W-shaped and spiral-shaped. Energy and Buildings. 2016;**133**:335-344

Chapter 7

Application of CFD to Prediction of Heat Exchanger Temperature and Indoor Airflow Control Simulation in Room Air Conditioner Development

Hajime Ikeda, Akinori Sakabe, Shingo Hamada, Mitsuhiro Shirota and Takashi Kobayashi

Abstract

In the development of room air conditioners, repeated experiments are conducted to improve product performance. In experiments, temperatures, and air velocities, however, the number and locations of sensors that can be installed in experiments are limited, and it is difficult to estimate the background of a phenomenon based on experimental data alone. Therefore, in design practice, we utilize quantitative verification by CFD in addition to experimental analysis. In this chapter, we describe two CFD models. The first model is a heat exchanger model of an indoor unit that has a cross-flow fan and a heat exchanger. This model is coupled with CFD model and one-dimensional refrigerant flow circuit model. This calculates airflow velocity, temperature, refrigerant flow pressure, and enthalpy. The second model is an indoor airflow model of a room. The room air conditioner can control the temperature and the velocity distribution of airflow in a room, which influences user thermal comfort by controlling the direction, temperature, and volume of airflow by using the indoor unit outlet. This model can predict airflow velocity and temperature in a room.

Keywords: HVAC, room air conditioner, heat exchanger, refrigeration circuit, indoor airflow

1. Introduction

Wall-mounted room air conditioners were invented in Japan and spread mainly as standard household air conditioners in East Asia. These are split-type air conditioners where the indoor and outdoor units are separate. The indoor unit uses a cross-flow fan, which reduces the depth and length of the unit and allows it to be mounted on a wall. The ability to install the air conditioner on the wall has the advantage of reducing the floor space required for installation. Design problems in indoor units are (1) the fin-and-tube heat exchanger design and (2) the airflow control. In the following sections, we introduce two numerical models for development of indoor units.

1.1 Fin-and-tube heat exchanger model

Figure 1a is a photo of an indoor unit **Figure 1b** illustrates a sectional view of the internal structure of the indoor unit. The indoor unit consists of three parts: the fin-and-tubes heat exchanger, the cross-flow fan, and the vanes. The internal airflow field pattern can be explained with a two-dimensional sectional view [1, 2]. The fan supplies air for the heat exchanger from the air inlet, the heat exchanger raises the air temperature, and then the vanes control the airflow direction, which enhance the air conditioner's thermal comfort and energy-saving performance.

The fin-and-tube heat exchanger consists of refrigerant tubes inserted into fins. Each tube can be connected with a U-shaped bent tube to form a refrigerant flow circuit with an inlet and outlet for the refrigerant flow (**Figure 1c**). A refrigerant flow circuit has a refrigerant inlet and outlet, and the single continuous tube connecting each is called the "path."

An important improvement for heat exchanger performance is to investigate the air and refrigerant flow in the heat exchanger, identify areas that are causing energy losses that do not contribute to heat exchange between the air and refrigerant flow, and improve those areas.

A main objective of the fin-and-tube heat exchanger model development is to quantify the refrigerant flow state such as pressure, enthalpy, and temperature in wall-mounted indoor unit heat exchangers, especially those units equipped with cross-flow fans.

In previous research papers [3, 4], the heat exchanger was modeled as a straightforward shape (**Figure 2a**). The velocity and direction of the airflow in the heat exchanger were assumed to be unchanged, so that the airflow volume distribution was given in the first row of the tube array and the second and third rows were simply the same as the distribution in the first row. These assumptions allow us to solve for enthalpy of refrigerant flow, pressure, mass flow rate, and heat transfer between the refrigerant flow and air, even when the heat exchanger has multiple paths connected by a distributor device. The previous study [5, 6] developed indoor unit CFD models, which can predict fin pressure loss and heat transfer of fins. These models do not include refrigerant flow calculation, therefore, the transition between the pressure and entropy values from the refrigerant pipe inlet to the outlet in the path design was unknown. The previous study [7] developed an inclined heat exchanger CFD model

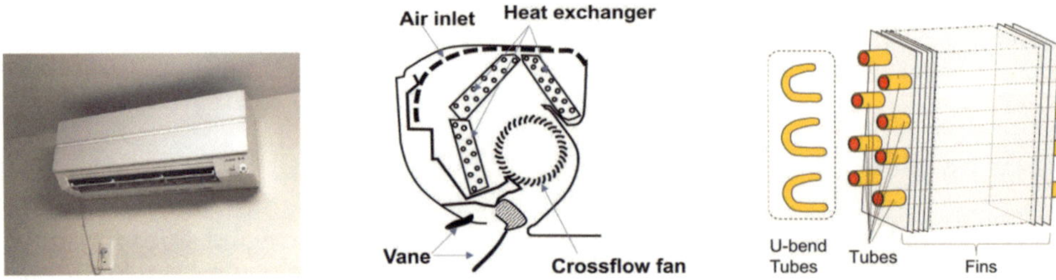

(a) (b) (c)

Figure 1.
Indoor unit: (a) a photo of an indoor unit, (b) internal structure of indoor unit, and (c) a structure of heat exchanger.

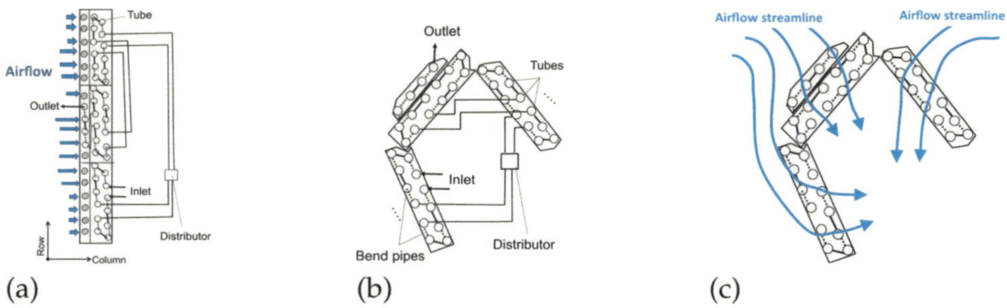

Figure 2.
Heat exchanger modeling: (a) a schematic of previous research model [3, 4], (b) an indoor unit of fin-and-tube heat exchanger, and (c) streamline of indoor unit heat exchanger.

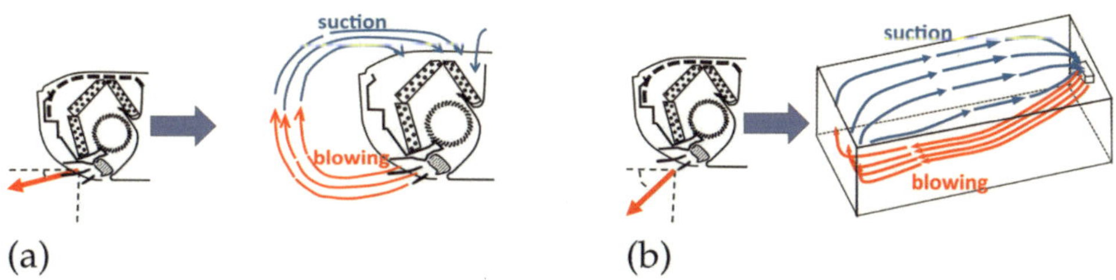

Figure 3.
Indoor airflow control: (a) short airflow circulation and (b) long airflow circulation.

and calculate airflow distribution. It incorporates air flow distribution to refrigerant circuit calculation model, however, this model can reflect indoor unit heat exchanger.

The heat exchanger actually used in the indoor unit is not a straight type but a folded type (**Figure 2b**) and airflow streamline is not a straight but curved form (**Figure 2c**). Since airflow does not flow straight through the first row of tube arrays, it is difficult to determine the air flow rate distribution input to the heat exchanger. Furthermore, the airflow volume distribution in the first row of the tube array is not the same as the tube array in the other rows of the actual heat exchanger, as assumed in previous research papers. Our numerical model is coupled with CFD for air-side calculations and with the one-dimensional refrigerant flow circuit model for refrigerant-side calculations. This CFD model reflects an airflow distribution in the fins and can calculate the pressure and entropy values of the refrigerant flow.

1.2 Indoor airflow model

Figure 3 shows an example of indoor airflow control. A wall-mounted air conditioner was installed on the wall and positioned to blow warm air from a high position in the room. The distance of the blowing air can be varied by controlling the fan speed, heat exchanger temperature, and vane angle. **Figure 3a** is a schematic diagram of short circulation control. The red line represents the airflow moving away from the indoor unit, and the blue line represents the airflow approaching the indoor unit. For explanation, we refer to the red line airflow as the blowing airflow and the blue line airflow as suction airflow. When the vanes are angled upward, the airflow is directed upward. In the heating mode, the temperature of blowing airflow is higher than the air temperature in the room. The blowing airflow floats upward because a buoyant force works. The airflow circulates near the indoor unit, and the reach of the airflow of the tip is short.

Figure 3b shows a schematic diagram of long circulation control. When the vane is angled down, the airflow is directed downward. Because the suction force is weak, blowing airflow tends to travel farther from the indoor unit and has a long circulatory path.

The airflow control is a technology that sends hot or cold air to various positions. To improve energy-saving performance and comfort, it is optimal to control airflow factors, for example, by avoiding airflow to areas of the room where the user is not present and selectively applying more airflow to body parts that contribute significantly to comfort, such as the feet [8].

If such airflow control can be simulated numerically, the work required for testing the actual equipment can be reduced. The objective of developing indoor airflow model is to replace testing with numerical calculations and to quantitatively evaluate wind velocity and temperature distributions, which are difficult to measure in testing.

Although there have been many studies using CFD in indoor airflow [9], few studies have dealt with split-type air conditioners. Casado et al. [10] verified the airflow temperature distribution using a split-type air conditioner. In this study, the vane angle was fixed. The case where the vane angle and wind direction are varied is verified by Lee S et al. [11]. The numerical model was validated using experimental data under isothermal conditions by varying the vane angle and wind direction. The numerical model was validated using experimental data under non-isothermal conditions with varying vane angles and airflow directions. Experimental data under non-isothermal conditions were used to validate the numerical model. Temperature distributions were obtained when the air conditioner was operated in heating mode and the vane angle and airflow rate were varied.

2. Fin and tube heat exchanger model

2.1 Numerical model

Our numerical model couples with CFD for air-side calculations and with the one-dimensional refrigerant circuit model for refrigerant-side calculations. The air-side model consist of two CFD models. The first model calculates airflow velocity and pressure by solving an unsteady Reynolds-averaged Navier–Stokes (RANS) equation model. The second model calculates airflow temperature and amount of heat exchange of each tube by using first model calculation result. This model solves a steady heat conductivity equation where the airflow velocity field calculation result is fixed. The refrigerant flow side model can calculate refrigerant flow pressure, enthalpy, and flow rates on each tube by using circuit model of a heat exchanger path. This model inputs heat of tubes from the air-side model result. **Table 1** denotes the input and output relationship of these models.

2.1.1 Air-side model

A commercial CFD software, SCRYU/Tetra (CRADLE 2018) [12], was used for air-side models. Two air-side CFD models, airflow model, and heat conductivity model use same CFD mesh. **Figure 4** shows the CFD mesh of the overall heat exchanger and the cross-flow fan, and **Figure 4b** shows the fin surface mesh. This mesh consists of a

Model	Method	Inputs	Outputs
Air-side model	CFD (airflow)	Fan rotate speed	Velocity and pressure
	CFD (heat conductivity)	Temperature of tubes	Heat of tubes
Refrigerant flow-side model	One-dimensional circuit model	heat of tubes	temperature of tubes

Table 1.
Fin-and-tube heat exchanger model.

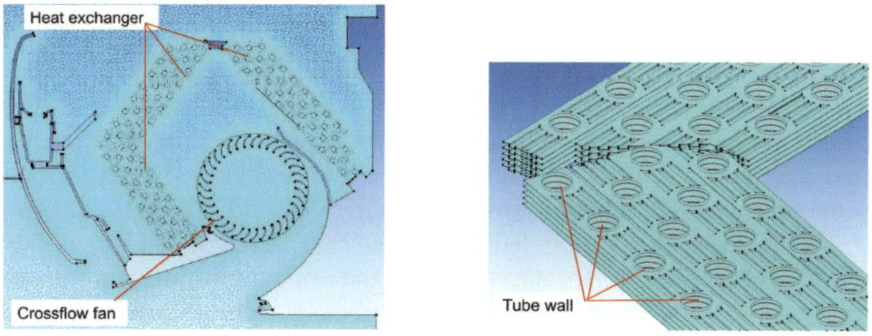

o (a) (b)

Figure 4.
Mesh of indoor unit: (a) internal of indoor unit and (b) fins and tubes. Tube walls are boundary condition regions of CFD (heat conductivity) model, which can be set to temperature from the result of refrigerant flow-side model.

stationary region mesh and a rotating region mesh. For rotating region, we created 0.25 mm to 0.50 mm tetrahedral meshes. Four prism meshes of 0.1 mm are inserted along the fan surface. Arbitrary Lagrangian Euler (ALE) method is adopted for fan rotating calculation. A tetrahedral unstructured mesh size was 0.25 mm to 50 mm and further prism layers were inserted along the walls. For fin walls, tetrahedral mesh size is 0.25 mm and 0.1 mm; two prism layers were inserted along the walls. Actual typical indoor unit heat exchanger width is 666 mmm and has about 555 fins. In order to limit the number of meshes and perform calculations efficiently within our computational resource, we designed our numerical model to have width of 6.0 mm and five fins. Previous research papers [2, 5] used two dimensional CFD model for simulating cross-flow fan airflow pattern. We used thin CFD model (6.0 mm width model) that models a the cross-sectional view at a representative position. For the thin width, this model airflow pattern is two dimensional. However, the actual indoor unit has filters and unit ribs that block the airflow at some positions. The cross-sectional view of the indoor unit has different shapes in different positions and, therefore, airflow patterns are different at these positions. When calculating the amount of heat exchange for each tube, the results are multiplied by 100.0, which is an adjusted constant to consider the influence of some different shapes of cross-sectional view of the actual unit.

2.1.2 Refrigerant flow-side model

We constructed a circuit model, which is made of node elements and tube elements to solve refrigerant flow state (**Figure 5a**). The node element preserves

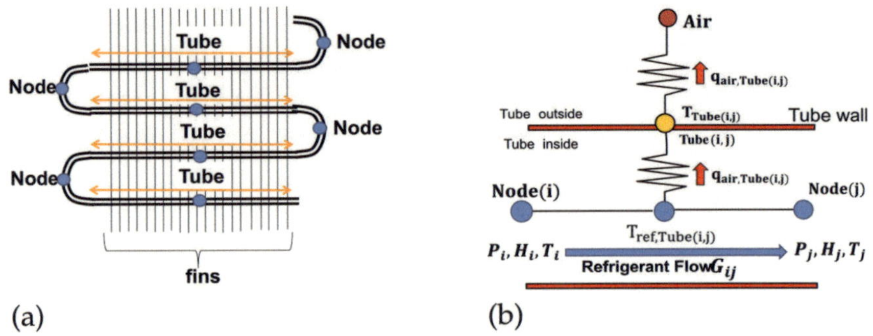

Figure 5.
Refrigerant flow circuit model: (a) arrangement of node and tube elements in the heat exchanger and (b) a relationship of the node and tube elements.

refrigerant flow temperature T (°C), enthalpy H (kJ/kg), and pressure P (Pa). The tube element is connected to two node elements. It preserves the tube length l (m), tube inner diameter ϕ (m), the refrigerant flow rate G (kg/s), the amount of heat transfer between refrigerant flow and the air q_{air} (W), and tube wall temperature T_{tube}. The distributor can be modeled with more than three node elements with tube elements. **Figure 5b** shows the relationship of the node and tube elements, indicated by the two nodes: Node (i) and Node (j), the tube elements connecting with Node (i) and Node (j) represented by Tube (ij).

Let P_i be the element pressure of Node (i), T_i be the temperature, and H_i be the enthalpy. The difference of Node (i) pressure P_i and Node (j) pressure P_j can be written by using the function $\Delta P\left(G_{ij}, \phi_{ij}, l_{ij}\right)$, which calculates pressure loss from the Tube (ij) refrigerant flow rate G_{ij} using Blasius equation and Lockhart-Martinelli correlation [3]. The relationship of P_i, P_j, and $\Delta P\left(G_{ij}, \phi_{ij}, l_{ij}\right)$ is as in Eq. (1).

$$P_i - P_j = \Delta P(G, \phi, l) \tag{1}$$

The relationship of the connecting node elements Node (i) and Node (j), enthalpy H_i, H_j, and G_{ij} is shown in Eq. (2).

$$H_i - H_j = q_{air,ij}/G_{i,j} \tag{2}$$

Suppose that the refrigeration circuit reaches the steady state, at any node element Node (j), the total refrigerant flow rate is equal to 0 (kg/s).

$$\sum_{j=1} G_{i,j} = 0 \tag{3}$$

Eq. (1) is the momentum conservation equation, Eq. (2) is the energy conservation equation, and Eq. (3) is the mass conservation equation.

Assuming that the number of node elements is n and the number of tube elements is m, the number of unknown variables is n each for P_i and H_i, and m for G_{ij}, for a total unknown variable is $2n + m$. The number of equations is n for each of Eqs. (2) and (3), and m for Eq. (3), for a total number of equations is $2n + m$. Since the number of unknown variables and the number of equations are equal, the value of the unknown

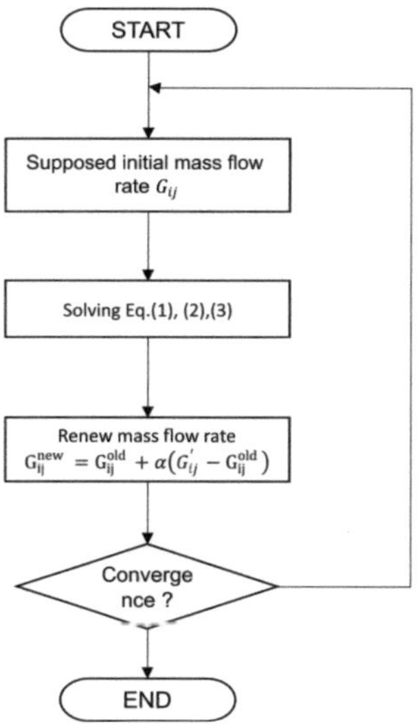

Figure 6.
Flowchart of refrigerant flow circuit model.

variable can be uniquely determined from equations Eq. (1), (2), and (3). **Figure 6** is a heat exchanger calculation flowchart. We can obtain refrigerant flow P_i (Pa) and enthalpy H_i kJ/kg for all node elements.

We can calculate temperature of node elements T_i from P_i and H_i by using refrigerant property database REFPROP [13]. We introduce the refrigerant flow temperature of $Tube_{ij}$ element as $T_{ref} = (T_i + T_j)/2$ (°C). We can then calculate wall temperature $T_{tube,ij}$ from the heat transfer rate of tube α_{ij} (W/m^{2}K), tube surface S_{ij} (m^2), tube wall thickness t (m), and tube wall heat conductivity (W/m^{2}K) as follows:

$$T_{tube,ij} = T_{ref} + \left(\frac{1}{\alpha_{ij}s_{ij}} + \frac{t}{\lambda}\right)^{-1} q_{air,ij} \tag{4}$$

2.1.3 Coupling calculation

Figure 7 is a flowchart of coupling calculation. Initially, the air-side CFD (airflow) is performed once. Then each tube wall temperature $T_{tube,ij}$ is assumed as the initial condition and CFD (heat conductivity) process is performed. This CFD (heat conductivity) process outputs each tube $q_{air,ij}$. Third, the refrigeration circuit simulation process calculates the refrigerant flow state using $q_{air,ij}$ and outputs $T_{tube,ij}$, which are different from the CFD initial condition $T_{tube,ij}$. The second and third process is an iterative loop processes.

In the previous research [5], it was necessary to fix the temperature of the tubes $T_{tube,ij}$i to perform the CFD (heat conductivity). In addition, it was not possible to calculate refrigerant enthalpy. In this coupled calculation, the heat transfer from the air to the refrigerant flow is the same value ($q_{air,ij}$) for both the air-side and refrigerant flow calculations. This makes it possible to calculate the refrigerant enthalpy at the same time the temperature of tubes is calculated iteratively.

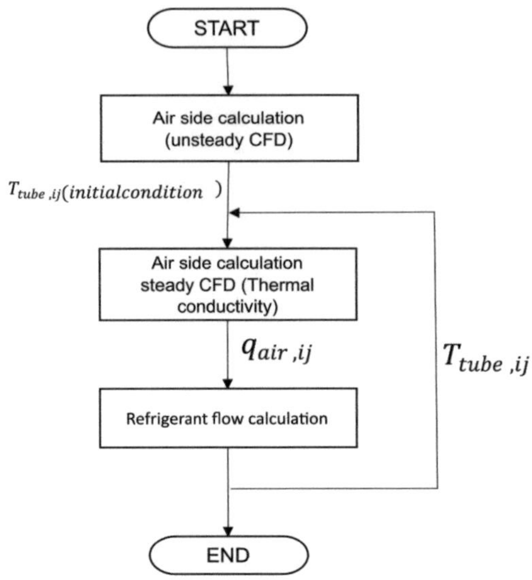

Figure 7.
Flowchart of coupled simulation.

	Fan rotation speed (rpm)	Volume rate (m^3/min)	Capacity (W)
Experimental	1555 (1.00)	22.2 (1.00)	8000 (1.00)
Numerical model	1550 (0.997)	21.6 (1.03)	8033 (1.04)

Table 2.
Air-side CFD result.

2.2 Result

The coupled calculations were verified using measured data from an actual room air conditioner in heating operation. The heat exchanger has six paths upstream of the refrigerant circuit, four paths in the midstream using a distributor, and two paths further downstream.

Table 2 is comparison of experimental and numerical data of air-side CFD model. Air flow volume rate is from the air-side CFD (airflow) result and heat capacity value is from the air-side CFD (heat conductivity). Amount of heat capacity of CFD is 4% larger than experimental.

Figure 8a and **b** show the air-side CFD (heat conductivity) result at the third run. **Figure 8a** is the airflow velocity contour plot, and **Figure 8b** is the airflow temperature contour plot. The value of the airflow velocity entering the heat exchanger is between 3.0 (m/s) and 4.0 (m/s) at the most area. There are areas where the airflow velocity is lower than 1.0 (m/s), such as at the front bended area and at the part of back of the heat exchanger. Previous research papers [3, 4] assume that the direction of airflow enters vertically into the front of the heat exchanger and direction are not changed, but calculation result shows airflow enter the front of the heat exchanger at various angles and that the direction changes inside the fins.

The air temperature distribution can be roughly classified into three areas. The area with a temperature near 20 °C is the airflow flowing into the inlet of the indoor unit.

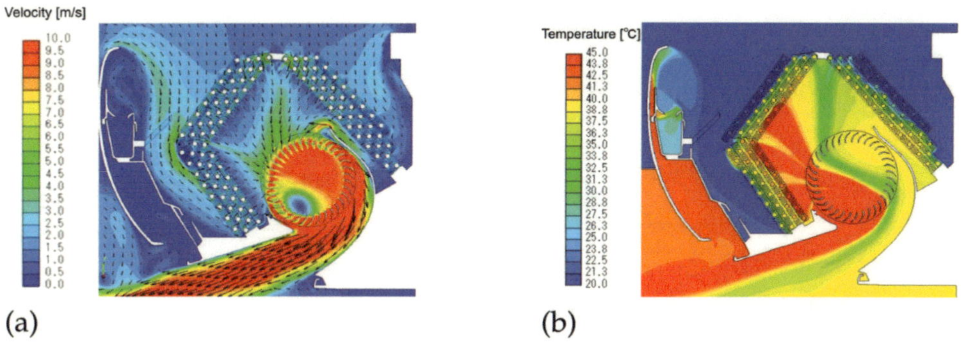

Figure 8.
Result of air-side CFD calculation: (a) airflow velocity contour plot; (b) airflow temperature contour plot.

Since the temperature of the refrigerant state is 20 °C in the liquid phase, the temperature of the airflow passing through the tubes at the top of the heat exchanger, which is in the liquid phase, is around 20 °C. The temperature of the refrigerant state in the two-phase flow is approximately 40 °C. The temperature of the airflow passing through the tubes in the two-phase flow section rises as it exchanges heat with the refrigerant. The tubes of the downwind side of the heat exchanger flow refrigerant vapor phase state, and the temperature of the pipes in the vapor phase exceeds 45 °C. The temperature of the airflow passing through these tubes is above 45 °C.

Figure 9 shows the fin surface temperature, focusing on the part of the fin where conductivity occurs between the vapor phase flow tubes (high temperature) and vapor/liquid phase tube (low temperature). A role of a heat exchanger is a heat exchange function between air and refrigerant. If the temperature difference in the refrigerant piping is large, heat exchange occurs between the refrigerant tubes. In this case, some tubes do not contribute to improved heat capacity, which is called "fin conductivity [4]." These models use fine fin mesh and can calculate heat transfer *via* fins.

Figure 10 shows the result of the refrigeration circuit model calculation using pressure and enthalpy plotted for node elements (p-h diagram). The refrigeration circuit consists of 12 paths. There are two refrigerant inlets, a distributor device, and four outlets. In the heating mode, the enthalpy of refrigerant flow decreases from the refrigerant flow inlet to the outlet because of heat transfer between the refrigerant and

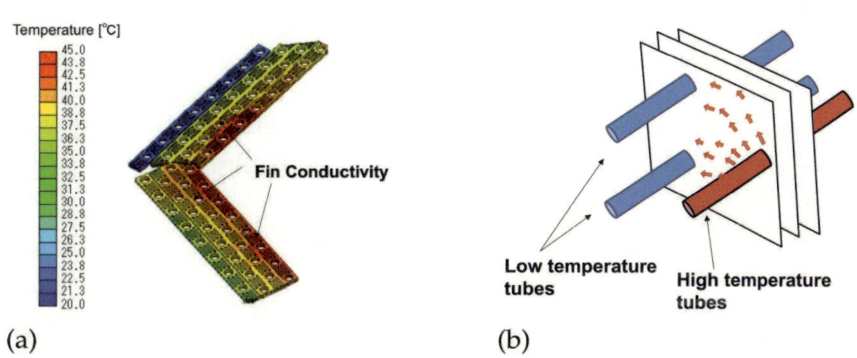

Figure 9.
Fin conductivity: (a) CFD result of fin surface temperature and (b) schematic illustration.

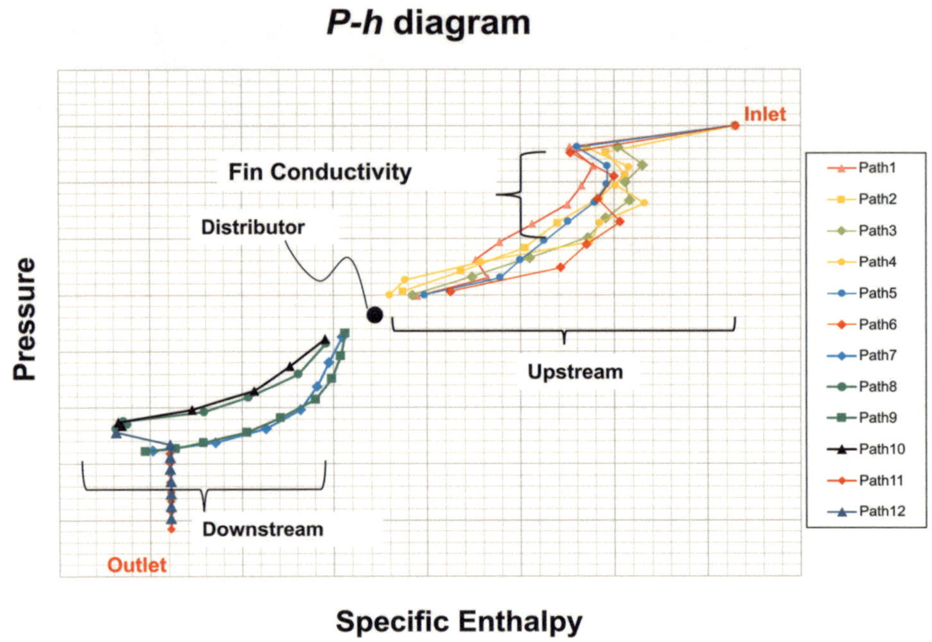

Figure 10.
A p-h diagram of refrigerant flow in the heat exchanger for each of 12 paths. Dots represent pressure and enthalpy values of node elements of tubes. Each path consists of 6 to 8 tubes. The distributor divides these paths into upstream side and downstream side. Paths 1 through 6 are upstream paths, and paths 11 through 12 are downstream paths.

the airflow. This result shows that our numerical model can quantify the energy loss due to fin conductivity.

3. Indoor airflow

3.1 Airflow temperature measurement test room

An airflow temperature measurement test room was used to develop the indoor airflow control model. This test room is constructed with wooden walls, windows, and doors that mimic a room in a Japanese wooden house. The outside air temperature in this room can be controlled by an air conditioning system. This room is equipped with thermocouples, which can measure indoor airflow temperature distribution. **Figure 11** shows the layout of this room and the arrangement of the thermocouples. The air temperature was acquired using a test room with thermocouples taken down from the ceiling. As shown on the right of **Figure 11**, the temperature distribution of the airflow can be visualized by interpolating the temperature in the space without thermocouples through heat conduction calculations.

3.2 Numerical model

3.2.1 Model development

Figure 12 shows how the mesh of the computational model is created. We simplified indoor unit cross-flow fans and vanes to an inlet boundary condition and indoor

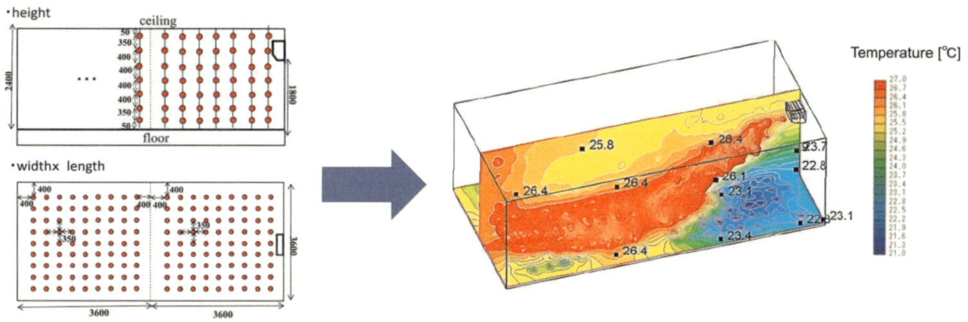

Figure 11.
Schematic diagram of the thermocouple arrangement. The test room is 3.6 (m) deep by 7.2 (m) high by 2.4 (m) wide. Thermocouples are suspended from the ceiling of the test chamber, and thermocouples are placed at intervals of 0.05 to 0.4 (m). A total of 1134 thermocouples were installed in a layout of nine thermocouples in the x direction, 18 in the y direction, and seven pointing in the z direction (height).

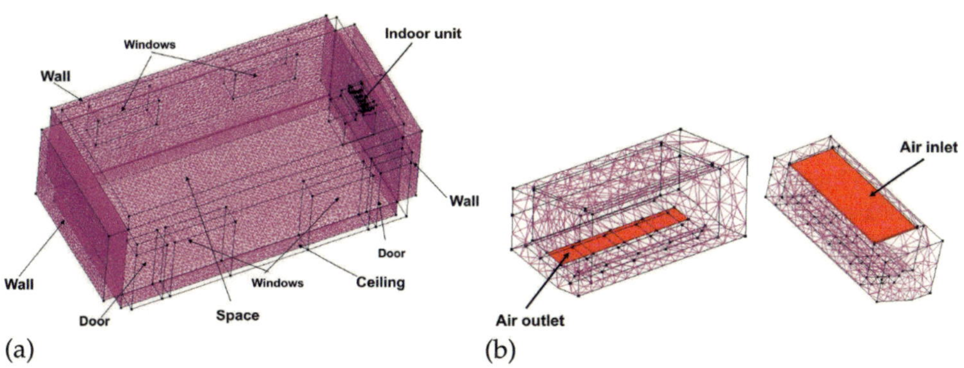

Figure 12.
CFD mesh. This mesh consists of a room space, walls, windows, and a door. The room space is the indoor airflow environment, and wall indicates the floor, ceiling, and front, back, left side, and right side walls. The room air conditioner was installed on a back side wall. Outer surfaces of walls can set outside temperature as the boundary condition.

unit inlet to outlet boundary condition. The boundary conditions for the inlet surface are (1) airflow volume rate, (2) airflow angle, and (3) airflow temperature. The walls, windows, and doors of the test room have small gaps through which drafts can enter and heat can leak out. It is difficult to model heat leakage from a draft numerically because air gap locations are unknown. Therefore, the initial temperature of the wall was set lower than the room temperature to model the heat leakage by the air gaps. The outer surface of the walls was set to the boundary condition of the outside air temperature in the experiment.

3.2.2 Buyoyant modeling

We use unsteady RANS model for indoor airflow control model. The governing equations were discretized by the finite volume method and solved using the SIMPLEC method. We used commercial CFD Code SCRYU/Tetra [12] for the calculation. To realize a rotating fan turbulence flow, the inlet condition was also given turbulence kinetic energy of 10 (m^2/s^2) and turbulence dissipation of turbulent kinetic energy of 0.1 (m^2/s^2). These values were obtained from the result of previous air-side

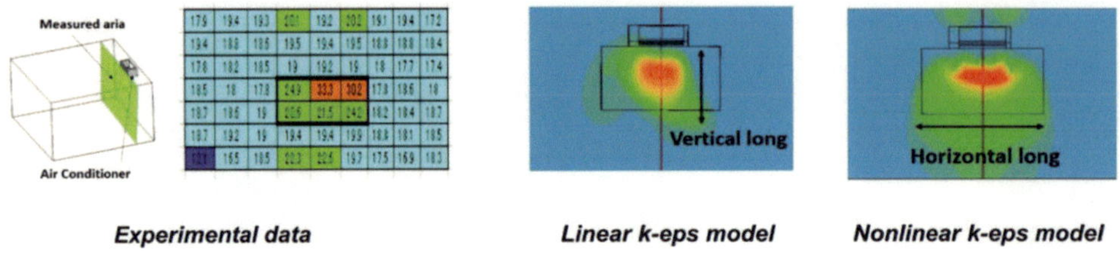

Figure 13.
Temperature distribution. We measured 54 points of air temperature at 0.4 (m) in front of the air conditioner and compared two RANS models.

CFD calculations. The blowing airflow from air outlet is acted as buoyant force, so that the Boussinesq approximation was used for buoyancy force as a momentum conservation equation.

To evaluate the influence of the turbulence model difference, we performed an experiment and two RANS model calculations. One model is linear RANS (standard k-epsilon model) and the other is used nonlinear RANS [14]. **Figure 13** shows the results of an experiment and two calculations. The experimental result and calculation of the nonlinear turbulence model for temperature distribution show the long warm air region in a horizontal direction. Otherwise, linear turbulence model shows a vertical long hot air region. The standard k-epsilon model is isotropic turbulence and the Reynolds stresses act equally in all three directions in three-dimensional space. Since buoyancy forces act only in the vertical direction, they cannot be reproduced by an isotropic model. Therefore, we consider that the anisotropic model is closer to the experimental temperature distribution than the isotropic model. We selected a nonlinear turbulence model for investigating the influence of the buoyancy force.

Table 3 is specification of CFD for our indoor air-control model.

Items	**Contents**
Turbulence model	Nonlinear RANS [14]
Buoyancy model	Boussinesq approximation
Scheme	SIMPLEC method
Time step	The time step is variable and the coolant number is fixed at 3.0.
Inlet condition	(1) Airflow volume rate, (2) airflow angle, and (3) airflow temperature are set for air-outlet region.
Outlet condition	Same value of airflow volume rate of inlet condition for the air inlet region.

Table 3.
Specification of CFD.

3.3 Experiment

We performed three experiments and calculations using the same control parameter conditions. **Table 4** lists the airflow control parameters of (1) angle (°), (2)

Items	Test (A)	Test (B)	Test (C)
Vane angle (°)	50	50	70
Volume rate (m^3/min	10.7	14.2	12.3
Temperature (°C)	47.2	42.0	45.5

Table 4.
Specification of CFD. Boundary condition for air outlet.

volume rate (m^3/min), and (3) temperature (°C). Angle is defined as the measurement between the horizontal line and the airflow direction.

3.4 Result

In the experiment of **Table 4**, tests (A) and (B) resulted in short airflow circulation. Test (C) resulted in long airflow circulation. We compared the tip of blowing airflow to verify the numerical results.

3.4.1 Short airflow circulation control

Figure 14 compares the airflow tips of tests (A) and (B) with the temperature contour plots at the center of the room. The calculated results are 180 seconds after the initial condition for both (A) and (B). The initial temperature of the indoor air was set to 24 °C and the initial temperature of the wall was 24 °C. Under the conditions of test (A), the blowing airflow floated in the center of the room and did not reach the floor in both experiments and values. However, the distance traveled by the tip of the blowing airflow was different (experimental, 3.6 m; numerical, 2.0 m). Under test (B) conditions, both experimental and numerical airflows contacted the floor; moreover, the distance of the tip of both experimental and numerical airflows was 3.6 (m).

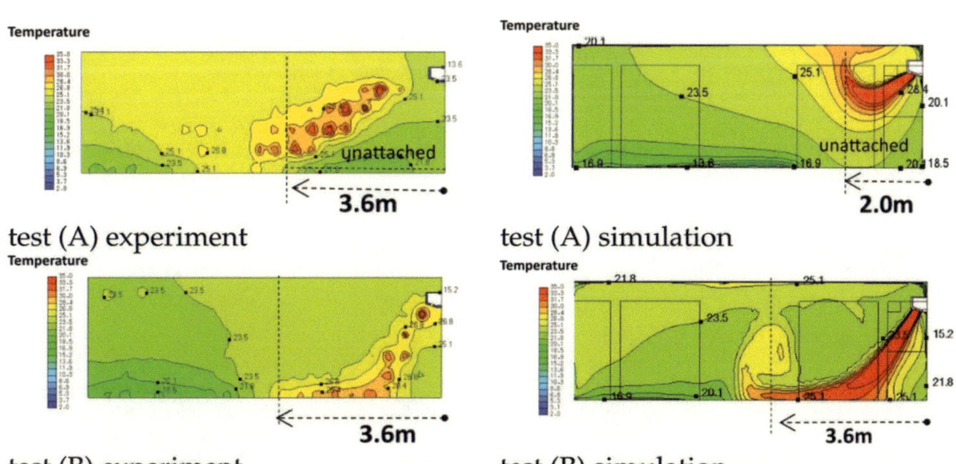

Figure 14.
Indoor airflow experiment and calculation. Comparison of test (A): Both experimental and numerical blowing warm airflow floated in the middle of the room, but the distance traveled by the tip of warm airflow was different. Comparison of test (B): Both experimental and numerical airflow touched the floor.

3.4.2 Long airflow circulation control

Figure 15 shows the results with test (C) conditions. **Figure 15a** visualized measured temperature by experimental thermocouples position. **Figure 15b** is the numerical simulation result of test (C) condition. The initial temperature of the room air was 20 °C, and the initial temperature of the walls was 10 °C. The contour plot is 180 seconds after the initial condition. Under the test (C) conditions, the both experimental and numerical results were such that the blowing airflow reached the floor and the tip of the blowing airflow reached the wall opposite where the room air conditioner was installed.

Figure 16 is three-dimensional isosurface of temperature at 26 °C. In the experiment, the temperature isosurface of 26 °C is limited to the floor, but in the calculation, it reaches the opposite side of the wall where the air conditioner is installed and the isosurface of 26 °C covers the half of the ceiling. In this model, the amount of heat leakage is simulated by using the initial temperature of the room walls as a parameter. This method can adjust the heat balance between the room air and the room walls, but it cannot simulate the airflow from air gaps. In test (C), blowing airflow travels along walls, such as floor and ceiling surfaces. The area where airflow flows along the wall surface in test (C) is larger than that of tests (A) and (B). It is possible that the airflow gaps on the wall surfaces in the test (C) are larger than in tests (A) and (B).

The positions of the arrival of the blowing airflow could be predicted by the calculation model, but the absolute values of the temperature distribution differed between the calculations and experiments.

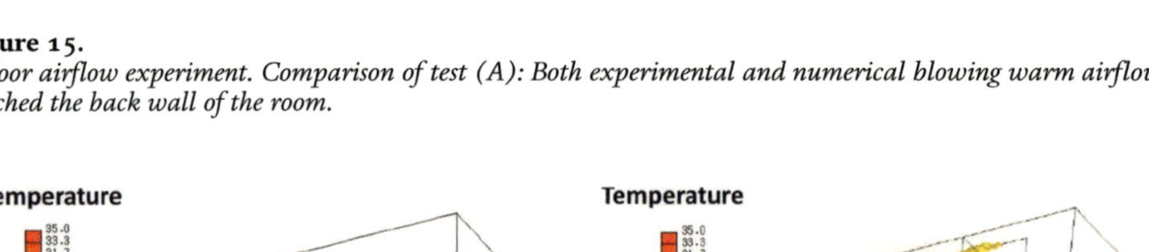

Figure 15.
Indoor airflow experiment. Comparison of test (A): Both experimental and numerical blowing warm airflow reached the back wall of the room.

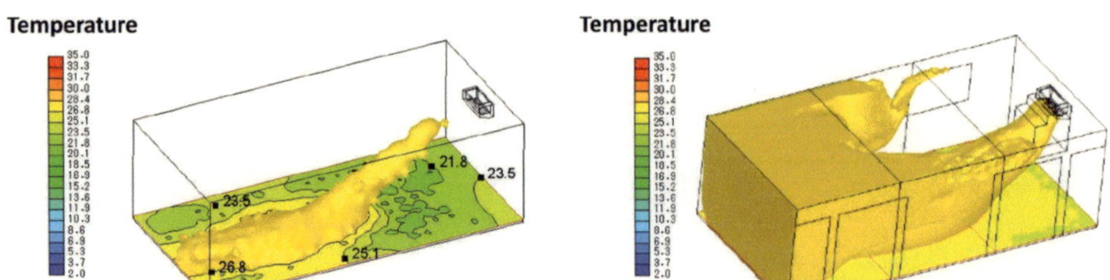

Figure 16.
Indoor airflow experiment and calculation of test (C). The temperature of isosurface is plotted at 26 °C.

When calculations are used as a surrogate model for experiments, it is necessary to analyze the calculation results, taking into account the differences in absolute values of temperature.

We found some differences in the experiments and calculations; however, qualitatively, the airflow calculations are similar to the experiments. The ability to compare experiments and calculations has enabled us to know the difference between experimental and calculated results. By knowing the difference, the number of experiments can be reduced by calculation, thereby reducing the cost of designing airflow control.

4. Conclusions

We described two numerical models for (1) a heat exchanger design and (2) an indoor airflow control. The fin-and-tube heat exchanger model consists of CFD and a one-dimensional refrigerant circuit model. CFD model simulates airflow velocity and temperature, and refrigerant circuit model calculates refrigerant flow pressure and enthalpy. These two models are coupled using the value of amount of heat transfer between air and refrigerant. The indoor airflow model can simulate indoor airflow temperature and velocity by the room air conditioner airflow control. We verified three simulations and the measured temperature data. The arrival position of the blowing air was found to be consistent between experiments and calculations, but some differences were found in the airflow temperature distribution in the room due to the existence of gaps in the wall.

Author details

Hajime Ikeda[1*], Akinori Sakabe[2], Shingo Hamada[2], Mitsuhiro Shirota[2] and Takashi Kobayashi[1]

1 Design System Engineering Center, Mitsubishi Electric Corporation, Tokyo, Japan

2 Shizuoka Works, Mitsubishi Electric Corporation, Tokyo, Japan

*Address all correspondence to: ikeda.hajime@da.mitsubishielectric.co.jp

References

[1] Kohya T. Historical Development of air conditioner technology (in Japanese). 2022. Available from: https://sts.kahaku.go.jp/diversity/document/system/pdf/099.pdf [Accessed 03 November, 2022]

[2] Toffolo A. On the theoretical link between design parameters and performance in cross-flow fans: A numerical and experimental study. Computers & Fluids. 2005;**34**(1):49-66

[3] Jiang H, Aute V, Radermacher R. CoilDesigner: A general-purpose simulation and design tool for air-to-refrigerant heat exchangers. International Journal of Refrigeration. 2006;**29**(4):601-610

[4] Kaga K, Kotoh S, Ogushi T. Prediction of heat exchanger capacity by thermal network method (influence of thermal conduction in fins on capacity of a condenser). Heat Transaction Asian Research. 2008;**37**(2):101-114

[5] Kudoh M, Sasaki S, Hatada T, Morimoto M. A simple method for predicting the performance of a heat exchanger mounted in an air conditioner (1st report, proposed reduced-mesh analysis modeling method). Heat Transaction Asian Research. 2004;**33**(1):12-23

[6] Liu N, Lai XL, Yan K, Zhang H. Investigation of flow and heat transfer characteristics on different heat exchangers of air conditioner. Applied Thermal Engineering. 2016;**103**:428-433

[7] Lee MS, Li Z, Ling J, Aute V. A CFD assisted segmented control volume based heat exchanger model for simulation of air-to-refrigerant heat exchanger with air flow mal-distribution. Applied Thermal Engineering. 2018;**131**:230-243

[8] Huizenga C, Zhang H, Arens E, Wang D. Skin and core temperature response to partial- and whole-body heating and cooling. Journal of Thermal Biology. 2004;**29**(7-8):549-558

[9] Nielsen PV. Fifty years of CFD for room air distribution. Building and Environment. 2015;**91**:78-90

[10] Casado AR, Hajdukiewicz M, de la Flor FS, Jara ER. Calibration methodology for CFD models of rooms and buildings with mechanical ventilation from experimental results. In: Ji G, Zhu J, editors. Computational Fluid Dynamics Simulations [Internet]. London: IntechOpen; 2020 [cited 2022 Nov 06]. Available from: https://www.intechopen.com/chapters/71460. DOI: 10.5772/intechopen.89848

[11] Lee S, Lee J, Kato S. Influence of vane angle on the effectiveness of air conditioning of wall-mounted split-type air conditioners in residential buildings. Science and Technology for the Built Environment. 2017;**23**(5):761-775

[12] Software Cradle Co., Ltd., "SCRYU/Tetra Users Guide"

[13] Huber ML, Lemmon EW, Bell IH, McLinden MO. The NIST REFPROP database for highly accurate properties of industrially important fluids. Industrial and Engineering Chemistry Research. 2022;**61**(42):15449-15472

[14] Batten P, Goldberg U, Chakravarthy S. LNS - An approach towards embedded LES. In: 40th AIAA Aerospace Sciences Meeting & Exhibit. American Institute of Aeronautics and Astronautics